BRITAIN'S TRADE AND AGRICULTURE

BRITAIN'S TRADE AND AGRICULTURE

THEIR RECENT EVOLUTION AND FUTURE DEVELOPMENT

by

MONTAGUE FORDHAM

WITH A PREFACE BY THE
EARL OF RADNOR

Routledge
Taylor & Francis Group

First published in 1932 by George Allen and Unwin Ltd.

This edition first published in 2018 by Routledge
2 Park Square, Milton Park, Abingdon, Oxon, OX14 4RN
and by Routledge
52 Vanderbilt Avenue, New York, NY 10017, USA

Routledge is an imprint of the Taylor & Francis Group, an informa business

© 1932 by Taylor and Francis

Publisher's Note
The publisher has gone to great lengths to ensure the quality of this reprint
but points out that some imperfections in the original copies may be
apparent.

Disclaimer
The publisher has made every effort to trace copyright holders and
welcomes correspondence from those they have been unable to contact.
A Library of Congress record exists under ISBN:

ISBN 13: 978-0-367-17896-3 (hbk)
ISBN 13: 978-0-367-17899-4 (pbk)
ISBN 13: 978-0-429-05834-9 (ebk)

BRITAIN'S TRADE AND
AGRICULTURE

BRITAIN'S TRADE AND AGRICULTURE

THEIR RECENT EVOLUTION AND FUTURE DEVELOPMENT

by

MONTAGUE FORDHAM
M.A.CANTAB.

Council Secretary, Rural Reconstruction Association
At one time a Director of Reconstruction in the White Russian
Province of Dróhicyn in Central Europe

WITH A PREFACE BY THE

EARL OF RADNOR

LONDON
GEORGE ALLEN & UNWIN LTD
MUSEUM STREET

FIRST PUBLISHED IN 1932

PRINTED IN GREAT BRITAIN BY
UNWIN BROTHERS LTD., WOKING

NOTE

The bracketed numbers ( ) in the text refer to notes in the Appendix, which contains incidentally much supplementary information of special interest to-day.

There may be errors in this Appendix, and indeed in the book itself—for no one is infallible. I invite correction and criticism.

PREFACE

The destruction of accepted views and recognized theories is often made use of to attract attention and pander to the very human weakness of a desire for a change. As a rule such efforts are accompanied neither by adequate reasons for discarding old ideas nor by constructive suggestions to take their place. Mr. Montague Fordham cannot be criticized from that point of view, for in his *Britain's Trade and Agriculture*, though he almost ruthlessly exposes the fallacies that lie behind the modern accepted views on industry and agriculture, he obviously writes without desire to attract attention, but simply to clarify facts and to suggest practical solutions. Moreover, throughout his book, he takes his views from life as it is, and not as it ought to be; he deals with current problems as they are to-day.

The first half of this book gives an analysis of our trade and agriculture, based not on the theories of economists such as Adam Smith, but always on actual experience of life. He describes, for instance, the methods of a pig-dealer's tout, which he considers are not uncharacteristic of our whole trading system. He shows how they are controlled not by economic laws, but by the simple fact that the tout probably has a wife and children to support and, anyhow, has to make a living.

The crisis through which this country (and incidentally

the whole world) is passing to-day is of such magnitude that the ordinary individual has almost given up trying to seek a solution. All that has been realized is that the old ideas are bankrupt and are no longer able to pay the debt of bringing the country out of its difficulties. Meanwhile those who have been appointed to lead us are finding out that such phrases as "The Law of Supply and Demand" and "Competition Lowers Prices" are flimsy structures and offer no shelter when we are driven into a corner by the realities of life. The author gives very full reasons why these and similar phrases do not help us, and, what is more important, he tells us what can help. Incidentally, in a section on stunts and slogans, he makes a brilliant attack on three prevalent, dominating methods of thought which create "pessimism," "optimism" and "internationalism."

Great emphasis is laid on the value of securing steady prices at a fair level. This is a subject on which the author is probably the first authority in this country. It is unfortunately one the importance of which has been obscured by the so-called Socialism of the legislative machinery involved. The disastrous result of vacillating prices is fully explained, with special reference to agriculture. Concurrently, reliance on development of our export trade as a means of restoring prosperity is disposed of by an analysis of what trade really is, and Revival of Agriculture is rightly, in my judgment, marked out as the more fruitful source of increased employment and increased national wealth. Thereafter the belief is put forward that, though it may be impossible for this country to lead any longer in trade, she is still able to lead in civilization.

It is probable that there may be many objections to the views put forward in this book on points of detail, but there will be very great difficulty in refuting the arguments put forward in support of the principles involved. They are drawn from actual experience.

In the course of comments on democracy the author quotes the case of a Conservative leader who, after the 1929 election, stated that for the first time in his life he found his audiences were really interested in and wanted facts. Much has occurred since that time, and audiences, who are the general public, now demand facts and are prepared to face them. It is perhaps too much to ask our politicians to purge their minds completely of catchwords and mass suggestions, the meaning of which they do not understand themselves; but if they have the courage to do so, they would certainly earn the respect of the bulk of the electorate, and by appeal to reason instead of sentiment would do a great good to their country. This opportunity now lies before them, and unless it is taken it may never occur again.

This book is simply and clearly written. It is full of amusing anecdotes taken from everyday life, illustrative of its arguments. This gives it a special character of its own, making it particularly suitable for those who wish to acquire knowledge easily and pleasantly.

I hope it will be read widely, partly because it contains a very close analysis of what the trade and agriculture of the country mean; but principally I wish it because it turns the mind into new channels of thought, leading to constructive action. At the moment the majority of English people are inclined to sit tight, waiting to see what happens. But the Englishman is essentially a man of action, and new ideas based on practical experience will lead him to action, and thus, it is to be hoped, ultimately to prosperity.

RADNOR

Longford Castle
    Salisbury

A FOREWORD ON THE
KALEIDOSCOPE AND PHILOSOPHERS

My friends often ask me what is the best introduction to the study of social problems.

There is a simple method of approach which has a special appeal to me and may have to others. Get a child's kaleidoscope—you can buy one for 10½d., one without a handle.

Now a kaleidoscope is a queer thing; it contains somewhere within it a number of pieces of coloured glass on which you look through a peep-hole. You see for a moment a definite brightly coloured pattern; but it changes with every movement, however slight, forming new patterns, themselves in their turn changing. There is always a pattern and never a permanent pattern. The social kaleidoscope is of the same character. There are the all the various coloured elements, psychological, political, financial, economic and so forth. They move, their relations change, and the pattern changes. An observer looks in through a peep-hole, sees a pattern and hurries off to write a book about it. Another follows and sees another pattern; a second book is put in hand, incidentally it proves that the first observer was mistaken. But meanwhile the pattern has changed once more and a new observer proves that the social system is quite different and that both his predecessors were wrong. Then perhaps the original observer returns, to find even another pattern; he writes a new book, and proves himself and everyone else wrong. And so it goes on. We have a multitude of counsellors but not wisdom.

If you want to understand the child's kaleidoscope, it is of little use merely to look through the peep-hole: you must take off the back and examine the mechanism. Then only will you understand it; moreover, you may, if you like, build up the pieces into a pattern and fix them in.

In the same way, if we wish to understand the social kaleidoscope, it is of little value merely to study the ever-changing pictures of social life, you must discover the mechanism: you must find out, not merely how life looks, but how life works.

Thus you may solve the riddle of civilisation. I suggest that this is not so difficult as it appears to those observers who, being specialists, are rarely familiar with life as it is.

This book deals with life as it is and is written for the general reader who may be advised to turn the pages at this point and start, at once, on the General Introduction.

But to those who are primarily interested in what may be called the intellectual background of life, I have also something to say. Read my book by all means, but you may also approach the subject from an entirely different outlook. That is, you may study, as I do, the underlying philosophy of life. I suggest to such thinkers that they should study Plato, Einstein and Bergson.

Study Plato, for it was Plato who realised clearly that from the very nature of things popular opinion was always, or almost always, mistaken; whilst he also taught through the mouth of Socrates how to detect errors and ascertain the truth. The Socratic method should put you on the road to the truth and should also, with the help of Freud and the psycho-analysts, make you immune against the stunts, slogans and illusions of the day—even, by the way, against the popular ones of the moment, which always appear incontrovertible.

Then study Einstein's explanations of Relativity and the relation of time to space and so to life: this helps you to see, in their true relation, all the various ever-moving elements that make up the social system and to realise the true relation of the past, the present and the future.

Finally, take Bergson's theory of Creative Evolution, which helps you to realise that a nation's fate is within its own hands, that a community has an element of free will, by the exercise of which it can and must, if civilisation is to survive, build up its own future.

It is these philosophers who have provided me with a basis for my methods of analysis and constructive thought.

MONTAGUE FORDHAM

The Severals, Seer Green,
    Beaconsfield
February 1932

CONTENTS

B

INTRODUCTION

THE GREAT FREE TRADE FALLACY

To understand, and thereafter to find the way out of the present chaotic conditions to be found in both trade and agriculture, it is as well to realise how these troubles arose.

To do this we must go back to the XVIIIth century—it was then that the tares were sown amongst the wheat that are now coming up to choke it. In that century, when the civilised world was faced with new problems, a strange wave of romanticism swept through Western Europe; we lost touch with reality and turned to phrases and fancies. Brilliant ideas detached from facts were welcomed and acclaimed as part of what was called "New Enlightenment." Speculative theory so named replaced the wisdom that came from tradition and experience. Words obtained a new power and value. Whilst popular phrases like the "Equal Rights of Man" and "The Sovereignty of the People" created the background for the French Revolution, it was the words "The Law of Nature" that made the appeal to the English intellectual leaders, and inspired the school of thought that ultimately created the present chaos. What was the meaning of "The Law of Nature"? It was, I suggest, in its origin, the very antithesis of the idea of a law of human justice derived from Christian teaching which had created the basis of much of the earlier economic system with its methods of regulated trade and just prices. Belief in "The Law of Nature," fortified by the *laissez-faire* idea, created a conviction that in the life of man the fight was and should be the controlling element, and that out of this fight civilisation would develop according to natural law and an equilibrium of Society be created.

The case was put forward thus: let us remove all restrictions and all communal control, let every man fight for himself and all will be well! Of course, the devil will take the hindmost; it is unfortunate, but it is part of "The Law of Nature." Besides, if the devil did not get the hindmost he might secure the foremost. The idea captured the minds of the English leaders and the principle of the trade struggle was adopted; ultimately the individual traders and financiers who won in the fight came up to the top and from that time controlled the government, and, what was even more important, secured great influence in both schools and universities.

The child that was born of "The Law of Nature" was christened "Free Trade."* There are few stories in history stranger than that of the great free trade fallacy. It is not a story, as some people suppose, of the power of an idea, but of the power of a phrase: a phrase which, if it had at any time any clear meaning, was always changing that meaning so as to describe or conceal a different line of policy. As a result, to-day, "Free Trade," if it has any meaning, means something quite different

* The history of the free trade movement is fully dealt with in that very remarkable book, *Protection and the Social Problem*, by Arthur J. Penty. To me this appears one of the most important books that have been written since the War. Mr. Penty is really, though perhaps he does not realise it, the Einstein of Sociology. He sees everything objectively and understands Relativity—the relations of things. One hopes that this will not be left to be discovered by future generations.

from "Free Trade" of a hundred years ago. One can indeed imagine an intelligent thinker born in the earlier years of the last century, whose life had been artificially prolonged to recent times, discussing the problem as it existed before the incidents of the last few years, in the light of the experiences of his early days. He might remark, using the phrase "Free Trade" in the sense of his youth, that "the United States of America have become a typical free trade country, for its internal trade now appears to be uncontrolled," and then adding, "Denmark, on the other hand, appears to have given up the free trade idea—in agriculture at any rate—and has introduced through her co-operative system the exact opposite, the policy of the opponents of free trade, a system of regulated trade." To which the up-to-date thinker might reply: "But we say that the United States is the leading example of an anti free trade country because they have tariffs and Denmark's boast is that hers is a free trade country because she has few or none."

The ancient gentleman might then have gone on to say that the change in the use of the word was remarkable, and might have added, if his memory were good, that the Act of Parliament passed in 1844 to secure "Free Trade" did not refer to tariffs at all; it was known as the "Dealer's Charter"* (1) and established internal free trade only. Tariffs, he had always thought, did not very much affect the idea. When he was a boy, many free traders were in favour of tariffs.

It is the power of the words and not of the idea, and

* (1) This and subsequent numbers refer to notes in the Appendix.

that only, that makes people say they are free traders. Not one person in a hundred can say even to-day exactly what "Free Trade" means to him; the more intelligent will probably explain that what they have in mind is that they are against import duties on the food we can produce at home and also, perhaps, on other products that we can manufacture ourselves.

From the very beginning, whatever their meaning, it was the words that won their way. How different would have been the course of civilisation if competitive trade had been described in the phrase "the Trade War," for that is what free trade really means and ultimately leads to. It is therefore fundamentally a double evil, for it is not only an evil in itself, but it creates the war spirit.

There is a war-time story which tells that when Professor Soddy, the distinguished scientist and sociologist, was asked the trite question, "Who made the War?" he did not reply "The Kaiser," the usual answer of the day, but "Adam Smith." The remark has a basis of truth in it, for it was Adam Smith and his followers who succeeded in popularising the belief in first "The Law of Nature," and then, under the shadow of this so-called law, in "Free Trade," the trade war. It was undoubtedly this glorification in the world of thought that created not only the atmosphere favourable to the Great War, but also led to the present industrial collapse. Free trade in its true sense of the trade war was accepted with certain limitations by almost every civilised nation, and the widespread character of the current economic collapse is due to this fact. Complete freedom of trade is of course sometimes modified by tariffs, but tariffs are but weapons in the war, sometimes weapons of attack, but more often of defence. Indeed, the indignation against tariffs was but the wrath of the orthodox against a weapon they deemed to be heterodox. "How disgraceful," said, in effect, the orthodox free traders, "to introduce tariffs into the struggle; we can never win against tariffs; our enemies are positively digging themselves in; we are for open warfare!"

But let us return to the history of the phrase and what went on behind the talk of politicians and orators that disguised its meaning. In the closing years of the XVIIIth century the anti free traders, who had been defeated earlier in their attempt to maintain what was called the "just" or "common price," or, as we should now call it, the "standard" or "guaranteed price," were still fighting for the enforcement of the common-law rule, that no one save the recognised retailers should deal in food at a profit. In this fight they still had the law courts behind them. The main free trade controversy had indeed at that time centred round the food problem and had almost dwindled down to one point: should dealers be allowed to traffic in food? The law said no, and fined and put dealers into prison. But free traders quoted Adam Smith, already to them almost a saint: "Dealers," he had said, "were no more to be feared than witchcraft." It is all a superstition, they argued. In the year 1800 they converted Pitt to their point of view, and as a result his government came over to the "free trade" side; the government then decided to make a leap in the dark, and from that time, presumably on its instructions, prosecutions of dealers became rare. Thus, without, so far as I can ascertain,

any special legislation, internal free trade—that is, freedom
of trade within tariff boundaries—succeeded in establishing
itself as a custom, if not the law of England. The free
traders had won. (2) It was not, however, until the year
1844 that the law was actually altered: meanwhile the
government appears to have connived at illegality. (3)

The real opponents of free trade, the believers in
regulated trade and standard prices, then gave up the
struggle, or else allied themselves with the so-called
"Protectionists" and ultimately disappeared: one hears
nothing more of them until the early years of the present
century.

The establishment in the year 1800 of free trade within
a tariff boundary was followed by forty years of disaster.
The normal effect of internal free trade asserted itself;
in our food supplies at any rate, the middleman, no longer
controlled, forced prices down against the producer and
up against the consumer. The price of bread, which
ceased to be regulated, went up to fantastic figures.
Agriculture was in great difficulties, but was kept going
with the help of heavy subsidies in the form of personal
and family allowances to labour. (4) Then came Cobden
and his school, who exploited the unfortunate position.
Let us go one step farther, was the case he put forward,
let us make a further leap in the dark: remove your tariff
boundaries and we can get back to the old idea; we shall
thus establish the equilibrium of Society, belief in which
was part of the free trade faith; as a result we shall stabilise
automatically the price of corn at an economic level. (5)
In the midst of this tragedy the people lost their heads,
but at last they believed Cobden's assertion that complete

free trade would stabilise prices: he got his way. But it was soon Cobden's turn to lose his head. In the early days of the controversy he had advocated free trade on the old lines, as an outcome of the "Law of Nature," the law of the fight. But his successes were too much for him, and he fell to the fate that defeats many leaders of men who happen to have no sense of humour; he believed himself to be an inspired messenger. "Free trade is one of the ordinances of the Creator, the law of the Almighty," he told the City of London in 1861. He no doubt believed it, and so carried conviction. This was long ago; it needed the Great War to awake suspicion of the cry "God is on our side," but in those days the idea was not suspect, and from that time onward the case for free trade was no longer based on the Law of Nature; it had become a divine, or at least a quasi-divine, precept.

Bright, rightly nicknamed "the Fighting Quaker," enforced Cobden's point, for with an assurance that almost amounted to blasphemy he attached his theories to the teaching of the Ten Commandments. "Though," he said on one occasion, referring to his beliefs in one of those wonderful phrases that carry conviction wherever they are heard, "though they were not given amidst the thunders of Sinai, they were none the less the commandments of God." (6) Those of us who can remember Bright, with his complete confidence and his marvellous voice cadences, can realise the effect that such declarations made. Here at last was a prophet. From that day almost to this, belief in free trade has ceased to be influenced by reason: it became an "Act of Faith" and its bitterest opponents have hardly dared to question the principles involved, evading the real issue by introducing their criticisms by some apologetic remark to the effect that free trade, like many other great moral principles, has not immediate practical application to the world in which we live and move and have our being. Perhaps, too, such worshippers feel a little like the old lady of whom it was noticed that whenever the name of Satan was mentioned in Church she made obeisance. An inquisitive niece once asked her reason. "It costs nothing to be polite, my dear," was her wise answer; "you never can tell where you may be going to: they do say, too, that Satan has many friends in this world."

Onward from that time, when Bright and Cobden discovered the quasi-divine character of free trade, until quite recent times, a quaint similarity appeared in the speeches and writings of the popular leaders of the English free trade movement. A little passion, a touch of piety, a belief that at long last all would be well, and a note of pride that the greatness of England was based on a free trade policy. It was true that all the promises of the free traders like Pitt and Cobden had failed to materialise, and had, indeed, created the opposite effects to those predicted; and it was also true that the evidence showed that the greatness of Britain might be due to other causes, such as the sudden expansion of industry and world development in which Britain happened to be able to take the lead. Truth was immaterial—faith and piety were the fundamental elements. It was even said that, on the occasions in the later years of the XIXth century when the free trade controversy was again raised, in obscure tabernacles prayers were offered up for its preservation.

Every aspect of the XIXth century controversy over trade has a strange fascination for the student of human thought. There was an orthodox creed and an orthodox opposition advanced respectively by the so-called "Free Traders" and "Protectionists" and the discussion was as a rule confined to one issue—not freedom of trade as against regulation of prices and distribution—but to the special point as to whether or not import duties should be introduced in the case of those articles sent from abroad that the nation could well produce itself. On the general desirability of the prosecution of the trade war, the two sides appear to have been substantially in agreement; they both wanted "to capture foreign markets"; it was only in the method that they disagreed. There was little real ground for defence of either method as a whole, and the controversy was inclined to take the form of proving that the theories of your opponents were wrong. That was not difficult, as both theories appear to be highly questionable; certainly they were not easy, perhaps impossible to defend by argument. It was all a typical political controversy—almost a sham-fight. The problem of import duties entirely dominated the situation and other points of view had no chance of expression; it did not even seem to occur to the nation at that time that there were other points of view and other methods of regulating trade worthy of consideration.

Financiers, traders and shippers who were primarily concerned in overseas trade supported free trade and opposed the introduction of import duties: moreover, investors who placed their money abroad became concerned in seeing that the British market was kept open for the food and other exports of the countries in which their money had accumulated; as a result they also appear to have passed over automatically to the other side. Powerful vested interests were thus created in support of the free trade idea.* The free traders also cleverly attached to their theory a belief that free trade served to provide us with cheap food and that the cheapest possible food was desirable.

It was in this connection that towards the end of the century some ingenious mind invented a new, and from the free traders' point of view, a priceless myth: "The Story of the Hungry Forties." This myth deserves explanation, for it had great force. As an outcome of the development of machinery and the various processes now called the rationalisation of industry, poverty and misery had been created amongst certain classes of industrial workers, whilst agricultural workers, who in 1835 (7)

---

* Viewing the problem from the outlook of the promotion of trade war, there is no real intellectual barrier between the point of view of the free trader and the protectionist, and it is quite easy for anyone to change from one side to the other. In fact, in the summer and early autumn of 1931 distinguished economists and politicians in this country were hastily changing; some declared themselves in favour of both systems, and one at least changed his opinion twice in a very short time.

Viewing the matter from another angle, it appears as if the different outlook of the protectionist and the free trader may depend on where his money is invested—at home or abroad—whilst the difference between on the one hand the free trader and his colleague, the protectionist—both of whom are at bottom advocates of the trade war—and on the other hand the anti free trader, advocating regulation of trade, is whether you want to make money for your class or to create wealth for your nation. I fear the point is subtle and may not be understood.

had been deprived of their family allowances, had been crushed down into a condition of terrible poverty. But the so-called "Hungry Forties" were not in Britain (8) a time of special poverty or of specially high prices of food. There was, in fact, only one year of dear food in the forties, when wheat went up to over 100s. a quarter and bread to about 1s. a four-pound loaf; that, curiously enough, was in 1847, substantially a free trade year, for during that year import duties were suspended or nominal. Excluding this year we find that the average price of wheat in the forties was about 55s. a quarter, with the four-pound loaf at about 8½d. in London and no doubt less in the country. These were not excessive figures for those days. In the fifties, under free trade, wheat went up to a far higher figure, and the average price of bread remained at approximately 8½d., which was also the average figure in the sixties. (9) There was then no connection between the troubles of those times and the Corn Laws. The real trouble was that so many people had not the money to buy bread; thus there was national tragedy. A sufficient interval having passed for the true facts to have been forgotten, a highly imaginative composer prepared a picturesque story of the disasters of that period. These troubles had arisen, the nation was told, out of the Corn Laws and the high price of bread. Old men were found who, perhaps for a pot of beer or other similar remuneration, searched their memories for stories of maybe the high-priced free trade year of 1847, or else of earlier days when, under a system of free trade within a tariff boundary, the prices of bread had gone up to fancy figures. These stories detached from their surroundings could easily be planted down in the forties. The story of the "Hungry Forties" thus created was still in use up to the General Election of 1931, and when referred to in pathetic terms by popular orators rarely failed to bring a sympathetic response from the audience.

Whilst the basis of the defence of free trade was its moral character, we find that from time to time in the world of more serious thought facts were being carefully selected, and arranged, to support the free traders' case. Adam Smith's School of Economics had told the country (10) that the wealth of nations depended on trade between nations—a thing valuable in itself and in fact an exchange of goods and services, and that theory, essentially unreal as it will be shown to be, formed a basis of many arguments and carefully selected illustrations. From the innumerable facts of life, illustrations can be found to support almost any theory. Ingenious arrangement, and even inventions of facts, went steadily on, and when towards the close of the last century the free traders' policy had gone far to destroy the agriculture it had been part of the programme of Cobden and earlier free traders to create, (11) a theory was put forward that the British Isles were unsuited to produce more than a small proportion of its own food, and that it was absolutely necessary to export goods to pay for the food we could not grow ourselves. (12) The statements taken together or separately are, as will be shown later, both untrue; nevertheless, on their adoption depended the reputation of the Liberals who were at the time the free trade party, for this theory was the only possible explanation for the failure of the party's promise that free trade would help agriculture. Accordingly, to secure their position, the politicians of that time not only stumped the country with this explanation, but somehow arranged that these specific statements should be taught in the new elementary state schools, where they are indeed to be found in the text-books of to-day. Liberals have always appreciated the importance of education, and have seen the value of training up the child in the way it should go. One has only to read the text-books used in schools and universities to note how skilfully a point of view has been introduced on both this and other questions.

It was perhaps the teaching received in his youth that accounts for the views of the Rt. Hon. J. H. Thomas, who, when he was Lord Privy Seal in 1929–30 in the Labour Government of that day in charge of the problem of unemployment, emphasised strongly the traditional Liberal views on agriculture and exports: indeed, he based some of his policy upon them. His point of view serves to illustrate the Labour Party's official attachment to free trade, to which, of course, there has been until recently, and may be again, strong opposition within the party's ranks. (13) This attachment to free trade, however, did not arise entirely from school teaching. The Fabians, and I believe the Independent Labour Party in their early days at any rate, thought it wise to conciliate Liberal opinion, and therefore bowed to the free trade theory, though it was fundamentally opposed to their socialistic teaching of control of national life by the community. Moreover, the older English Labour leaders—most of whom were ex-Liberals and brought their theories with them—particularly prided themselves on being intelligent, and the Liberals had somehow managed to create the impression that intelligence was on the free trade side. Finally, the erudite Fabian, Lord Passfield, who in the days when he was Mr. Sidney Webb was looked upon as a mine of accurate knowledge on all problems economic, social and political, was on the free trade side. (14) The extremists of the Labour movement have also always been attached to the free trade idea. Had not Karl Marx been a keen supporter of freedom for international trade? It had a real appeal to him, for he foresaw that, if persisted in, it would destroy our present civilisation. (15) This the Soviet Republic, following him, have also seen, and have devised accordingly.

Belief in free trade, inspired by hysteria, remained, almost up to the present day, an "act of faith." Recognising it to be a form of hysteria and, further, realising that politicians are peculiarly subject to hysteria, it is not surprising that the political devotees, the dancing dervishes, entirely lost their heads. In August of 1927 Mr. Philip Snowden* (16), the protagonist of the modern movement, was telling the Labour Party that it was necessary to increase the purchasing power of the peoples of India, China and Africa if we were to restore prosperity to our manufacturers. No doubt he was vaguely recalling the oft-repeated tales of how missionaries in Africa had been followed by traders, with the success of the latter in persuading the natives to wear clothes and drink gin, from which such extraordinary advantages had accrued to British trade. Three years later, in October 1930, Mr. Snowden was elaborating the idea. (17) If only, he

* Now Viscount Snowden of Ickornshaw.

suggested to a Lancashire audience, we could persuade
the Chinese to lengthen their shirts by a few inches,
all the looms of Lancashire would be at work. An unkind
critic pointed out that the Chinese did not wear shirts.
But what of that—the argument is clear; the line of policy
that ought to have been adopted by his party, then in
power, was even clearer. Obviously they should have
attached a labour orator to their special trade mission to
China, (18) to use, with the help of an interpreter, his
eloquence to teach the men and women of China to wear
shirts. He might, if he could have successfully evaded the
attentions of brigands, have stumped the country. "Wear
Shirts" might have been his slogan; "Day Shirts and
Night Shirts: Red Shirts for Bolsheviks, Black Shirts
for Loyalists." And all would be well. The "drummers,"
following the salvationer, would have taken the orders—
Lancashire would have been saved; unless, indeed, the
Japanese and Chinese traders had got in first.

There is a danger that the reading of the preceding para-
graphs may create a special prejudice against politicians
of the Victorian school. It is important to free your mind
from such prejudices, and to recognise that these de-
scriptions contain what is in fact merely a picture of
the political method of creating mass illusions, which is
a normal feature of political life even to-day. It is more
important to secure a wider outlook, which I hope to
suggest by an illustration that makes a special appeal to
me and may therefore do so to my readers. There is a
quaint resemblance between the crusade initiated by
Martin Luther and that of the free traders in its later

form. Martin Luther pinned his faith on "The Book"—it was "The Word" that counted. The free traders also idolised words. In both cases it was the worship of what in the East is called a *mantra*—a holy phrase. The two movements were also both heretical, and both in their early stages represented a minority. Nevertheless, when in each case the minority became a majority and secured power, their opponents were bitterly attacked as the enemies of their idol. But bibliolatry is once more completely discredited; very few believe to-day in the verbal inspiration of the Bible. Free trade is going the same way: it is knowledge of the facts that will destroy it.

We may now turn from history and philosophy to facts. Thereafter we can indicate shortly a possible way out of present troubles.

PART I

TRADE IN THEORY AND FACT

TRADE IN THEORY AND FACT

1. INTRODUCTORY

The wealth of nations, we have been and are indeed still told by popular expounders of the orthodox English economic theories on which our national policy has been based for nearly a century, depends on trade development. Trade explained as an exchange of goods and services is a process admirable in itself. Such was, and to some extent still is, the theory; unfortunately, like most of the theories of that particular school of thought, it has turned out to be untrue.

It will be useful—for the benefit of those who have no experience of trade—to examine these theories and try to give some picture of what trade really is. The word "trade" conceals a large variety of transactions which take place in the main between and through individuals. Talk about international trade has greatly confused the issue, for in the past, if we except the purchase of armaments and special transactions that have taken place in time of war, there have been very few transactions that could be described as trade by nations. It is, of course, true that in the last few years we find official or semi-official organisations representing governments engaged in business: for example, in Russia, where all overseas trade appears to be in the hands of the government; in Norway, where the importation and distribution of grain and grain products is managed by a Board elected by Parliament; in the United States of America, where the Federal Farm Board buys and sells agricultural produce; and in Brazil, where the coffee trade is under government control. But these special cases, though they may be the models for a new method, do not affect the general rule that trade has been in the past, and still is, an affair of individuals and associations of individuals rather than of nations.

Trade may be classified in various ways: on the one hand there is what may be called "constructive" or "Complementary Trade," creating wealth and employment; on the other, there is "destructive trade," creating poverty, misery and unemployment. Analysing trade from a different point of view, there is the form that may be called "Barter Trade," a real exchange of goods and services, whilst there are various forms of trade based on financial considerations—for example, what is called "Tribute Trade," a phrase which describes a group of transactions under which goods and services come from the peoples of one nation to those of another to pay their interest and dividends on investments, and there is no return of goods. The various forms of trade based on finance may be grouped under the general title of "finance trade."

It is quite easy to understand all these forms of trade if the explanations which follow are studied.

Trade, viewed from the point of view of the producers and consumers and the population as a whole, is therefore good, bad or indifferent; but from the point of view of the financier and the trader, trade is always or almost always good, for on all or almost all trade transactions there are immediate profits for the trader and the financier.

No one knows the whole truth about trade. It is only the bankers and financiers who might be able to tell us that, for the financial arrangements of practically every transaction pass through their hands; but they tell us little or nothing. Nevertheless, we do know enough about trade to be able to give a popular, and at the same time not inaccurate, picture of what really happens. Moreover, this is not intended to be a learned disquisition on trade economics. I will therefore take simple examples—some from my own experience—that will make the main facts clear.

2. OVERSEAS TRADE

We will take first overseas trade, not because it differs in principle from home trade, but because it has been so much talked about that it is easier to understand: we will begin with barter trade.

(a) BARTER TRADE

Barter still goes on occasionally in its primitive form on all sorts of scales, and there is talk of its revival as an ideal method. For example, the Brazilian government in 1931 bartered 1,050,000 bags of coffee with the Federal Farm Board of the United States for 25,000,000 bushels of wheat. (19)

But in overseas barter trade the interchange is through finance, and it is with this form of trade that I now propose to deal.

Nearly forty years ago, when I was in Ceylon, firms in Manchester were sending out cotton goods to be made into garments for the Cingalese. These goods were so beautiful that they had not, in those days, any possible sale in England and were so lasting that to this day pieces that I purchased in Ceylon are being used as draw-curtains in the windows of my cottage. In return Ceylon sent tea, which we could not have produced ourselves. This was, I judge, beneficial trade—advantageous to both countries: it belonged to the category called complementary trade, each community was exchanging what the other could not produce for itself. It can be criticised, it is true, on certain grounds, but on the whole it is defensible and benefited the peoples of both nations. This was the special form of trade that Adam Smith and his followers had in their minds when they described trade an an exchange of goods and services, beneficial to both nations. It is the only form of trade that has distinct economic advantages: in the early days of undeveloped commerce it was, perhaps, the usual form of overseas trade; but now it is less common. But there are many other transactions that lie on the border-line. If, for example, we send coal to Sweden and they send us back timber in exchange, it would be of the nature of complementary trade, and would for the moment be beneficial to the peoples of both countries. For to-day Sweden needs coal and we need timber. But such a transaction will at once suggest lessons for the future. We ought to conserve our coal—we cannot always live on capital—and coal is our capital. (20) Moreover, Sweden may not always need our coal. The

Swedes may either decide to buy elsewhere or to rely on electricity for power and heat. (21)

These two examples make clear what is meant by complementary trade. Some student of Economics should find out and tell us in a simple form how much of our trade is complementary. Such an investigation should start by studying our imports. There are certain things that, if we are to have them at all, we must get entirely from abroad. Tea and coffee, raw cotton, rubber, oranges, almonds, raisins and some other fruits and wine. We cannot produce more than a very small part of our tobacco. We no doubt want for electrical equipment much more copper than we can find in our own country, and some potash for fertilising the soil; also oil, so far as it cannot be produced from coal, is needed from abroad, and probably certain sorts of iron ore and other metals. We can hardly expect to produce all our own sugar. It is also always possible that even under the most favourable circumstances we should find it difficult to feed ourselves entirely and we may therefore be obliged to import a certain amount of wheat and other grain, and some part of our beef and mutton. In the final analysis, if we developed our natural resources, there would not, I think, be a great deal to import into this country; for Britain, even with its present population, is naturally a country that can be largely self-supporting in the essential things of life. Probably about half our present imports, i.e. food and manufactured goods of a total value of £500,000,000, could be advantageously produced at home, giving employment to about two million of our unemployed with corresponding increase in our national wealth. (22)

Moreover, it may be noted that all or almost all our needed imports can be obtained from our own dominions. The effect of such a development of home trade is discussed later.

Our student enquirer, having got these facts together, should similarly analyse our exports and find out how far they are needed by the countries that receive them: this done, we should have a clear idea of how much of our trade is really complementary and so beneficial.

Now let us take examples of the sort of trade that is dangerous and probably destructive in its effects on our civilisation, keeping still to barter trade. When the Rt. Hon. J. H. Thomas was grappling in 1929–30 with the problem of unemployment as Lord Privy Seal in the Labour government, he propounded a scheme for sending coal to Canada, to be paid for in increased importation of corn. Had his policy been adopted, the result would have been that we on the one hand, as in the Swedish case, would have been using up our coal—living on capital—whilst in return we should have been getting something that we might well produce ourselves. Such a deal can be compared to employing a man to empty your coal cellar, in order that the coal may be sold to purchase food that the worker might be occupied in growing in your garden. The transaction would not only impoverish this country, but might possibly put even more workers out of employ in agriculture and related industries than the number that were brought into employ in coal-mining; it could be of little, if any, real permanent advantage to Britain. But an even worse result might ultimately follow. The exportation of coal to Canada would be in a competitive market, and so would have all the worst evils of the free trade system. Such trade is a game of beggar-my-neighbour; first the peoples of one nation win and then those of another. If Mr. Thomas had had his way we should have probably soon lost in this struggle, and Canada might have bought her coal elsewhere. But even if we had won in the struggle, it would have been at irreparable loss; our victory would either have been secured by a further rationalisation of the coal-mines, which would have meant unemployment, or by reduction of wages, which would have meant that the miners had less money to spend, and so there might be unemployment in other industries. In the meanwhile land would have gone out of cultivation, or its productive power been reduced, whilst men dismissed from agriculture would have drifted away and could not have been replaced save by a special effort. It may also be well to consider what might have happened if Mr. Snowden had had his way and we had really succeeded in increasing our exportation of cotton from Lancashire to China. In this case the cotton would have to be paid for in some way. In so far as the payment was made in tea, it might easily dislocate our trade with India and Ceylon, reducing their exports to us, and so our corresponding exports to them. Indeed, whatever was the immediate effect on Lancashire in the increase of exports of cotton, the transaction would be likely to cause dislocation in other branches of industry; whilst finally the trade might well be killed by the competition of Japan (23) and the numerous Japanese cotton mills now established in China, or by the action of the Chinese themselves, who are concerned with setting up their own factories. Whatever may have resulted in the past from such transactions—and a large part of our overseas trade has in earlier days been of this character—over-development of export trade to-day, though it may have a temporary value, will be certain ultimately to result in unemployment and disaster.

Mr. Thomas's proposal for the exchange of coal for wheat stands for destructive trade in a simple form. There are other forms of trade that are far worse. In the latter years of the last century, when I was in Japan, leading statesmen were explaining their policy for the development of industrialism and trade. I advised them against it and pointed out the evils that had already occurred in our own country. They frankly recognised the evils, but said they were necessary evils, and would lead to the creation of the wealth needed to build up a new civilisation. They talked of progress; they were obsessed with Victorian ideas. Later on, I understood they were obtaining the necessary machinery from England, and probably from elsewhere; this export of machinery was a normal trade transaction, and the manufacturers who provided the machinery and equipment flourished. Concurrently we were helping them to build battleships to protect their trade. A firm of which I was a director got a small share of these battleship orders. From these orders British manufacturers benefited. But the ultimate results were disastrous, for Japan entered into competition with us from her own country, and later from her cotton mills in China, and has so helped to destroy our markets. No one can suggest that the Japanese or Chinese workers benefited from this process. Factories

in Japan, I was told, at one time were worked continuously night and day, week-days and Sundays, with one
eight-hour stoppage now and again when the machinery
was overhauled. The workers, drawn from the villages
and half-enslaved, were replaced as they broke down by
new recruits from a teeming rural population. The people
suffered, but, as usual in trade development, the traders
and financiers benefited. Finally, as a result of general
dislocation of trade, came unemployment, even in Japan;
whilst in England the end of it has been that English
workers have lost their jobs and unemployment in
Lancashire has become widespread. Displaced British
workers might, it should be noted, have been gradually
turned back to agriculture, to which so many of them
naturally belonged, had not our free trade policy already
prevented the development of agriculture.

We can go all over the world and find how the trade
of making machines to equip other countries has had
destructive effects. Jute factories built in Calcutta have
created unemployment in Dundee; cotton factories in
Bombay and Madras have created unemployment in
Lancashire; boot and shoe factories in Czechoslovakia
have created unemployment in Northampton and other
centres; linen factories in Russia have created unemployment in Belfast. "We all buy Russian linen to-day," a
lady said to me on one occasion, "it is so much cheaper."
No doubt it was, to her, but her gains are the nation's
loss; we have to pay in unemployment. (24)

But even this trade in making machinery to equip
other countries cannot be continued indefinitely, and it
should be observed that English firms were, in 1931,

receiving orders from Russia for the equipment and machinery which will enable the Russians to make their own machinery. Everybody seemed pleased when we got these particular orders! These various Russian transactions were, or will be, no doubt paid for in timber, corn and other food products, or perhaps oil. The provision of timber, under conditions, if contemporaneous reports in the Press were true, horrible in themselves, goes some way to destroy our complementary trade with Sweden, Norway and Finland. There is something to be said in support of the importation of oil from Russia, but the importation of corn, (25) however you look at it, is a disaster. It not only dislocated our trade with Canada and elsewhere, but was one of the causes of the fall in the price of wheat, that brought disaster on the farmers of East Anglia and created unemployment amongst thousands of workers. The same results occur from the importation of other food products. When you add to the price of wheat and other forms of food coming in from other countries the expenses of the relative unemployment, you realise how costly imported food may be.

Importation of corn from Russia is no new matter. I remember when I was a boy a cousin of mine, a corn merchant, coming into Royston market with a sample of Russian oats. "Hullo," cried the farmers, "here's Russian Oats!" It was said that he was booed out of every market in the countryside, under this nickname of "Russian Oats."

The tragedy that took place in East Anglia in those years, when people of all classes—squires, farmers, smallholders, blacksmiths, carpenters and labourers—were deserting the villages for the towns and colonies, remains as a vivid picture on my mind. It was later in life that I heard of what had happened in Russia. According to the account given me by a Russian authority, the corn taken in the days before the War from the Russian peasants resulted in the reduction of large districts to relative poverty, and in some cases aggravated the famines in which the people died wholesale. Free traders may suggest that the Russian peasants were paid for the corn—that they got something in return. It is more likely that when financiers, shippers and traders had taken their share, most of what was left went to the Jewish moneylenders, whilst any balance would go to the government in taxes. I am sure that anyone who has been, as I have, in touch with the life of the Russian peasants, will agree with me that this is probable.

We may stop for a moment to compare this transaction with the picture drawn by the orthodox economist of trade as an exchange of goods and services, beneficial to all concerned. There was, indeed, something in the transaction that might be called an exchange of goods and services, but whoever benefited, it was not the peasant producer, who got little or nothing. The economist may argue that this is a special case, a mere curiosity in trade; he would be wrong, for it is a commonplace of some branches of the food trade, to which I shall refer later, that the producer, on occasions, gets nothing, or next to nothing, and that though the price to the consumer may even be high, everything is absorbed by the financiers and the transporters and other middlemen.

We will return to our own country. The tragedy that

had occurred in corn-growing districts nearly half a century ago, from the importation of cheap corn, made a profound effect on the minds of the people, but they found their own remedies. Some fled to the colonies; others to the town and industrial centres, to replace the weaker city workers and create poverty, misery, slums and overcrowding. But even these remedies are denied to men to-day; for where is the place to which they can flee? Others remained in the country and turned to the production of fruit and vegetables, to milk and dairy produce or to fowls and eggs, in which, for some time, they secured a certain measure of success. But the success was only temporary: many growers of fruit and vegetables have in recent years been ruined or half-ruined, and in 1931 the profits of dairy farmers were dwindling and the position of the poultry farmers becoming precarious. Indeed, every, or almost every, branch of agriculture, has been on occasions, and may be again, undermined in turn as a result of the unregulated importations that our free trade system permits.

I will give one other example of barter trade that is illuminating. If you happen to have visited Holland in recent years you may have seen many fine herds of cows. You would then have congratulated the Dutch on their wealth and enterprise. But if you had enquired closely you might have learned that some, at least, of these herds were owned, or at least controlled, by a company with their headquarters in London, and that the milk was not consumed in Holland, but came over to this country either as tinned milk, or in special glass-containers to be sold as fresh milk. From that, I learned, arose a curious transaction. A Board of Directors, sitting in London, fixed the price of this imported milk—not, it may be noted, in accordance with the so-called "law of supply and demand," (26) for the price was fixed from time to time to cut in just below the English prices, to undermine the English producer and to prevent the development of home productions. And at times, there is reason to believe, the Board were prepared to sell their milk at a loss; they were concerned with controlling the trade at any cost.

There are two points in this little story worth noting; the actual ownership of the stock and the price-fixing. They can be illustrated by other examples. I will give one at once. I remember a day in the year 1922—the last day of my stay in the White Russian provinces of East Poland, where I had been concerned in the organisation of agriculture after the War. I rode round my district, which centred in the little town of Dróhicyn. Coming to a remote village called Popina, famous locally as the centre of the Battle of Popina between the Russians and the Germans, I found a herd of cows. It was an impoverished district, four armies had passed over it—Russian, German, Bolshevik and Polish. The Soltys, or headman of the village, came out to meet me. "Pan Soltys," I said, "you are rich men in Popina; you have cows." "Not so, Barín," was his answer. "They are not our cows; they belong to the Jews of Dróhicn; they own all."

The second point to be noted—the presence of a Committee sitting in London to fix prices to undermine English prices—is to be found also in the case of the Danish bacon trade, of which more will be said later.

We will now leave barter trade and consider certain trade transactions that depend on finance. Finance appears to control trade in many ways, but I propose to concentrate attention on two aspects of finance-controlled trade, which I will class under the titles of "Equipment Trade" and "Tribute Trade."

(b) EQUIPMENT TRADE

Since about the middle of the last century we have been busily engaged in fitting up other countries—the United States, the Argentine, India, Canada, Australia and New Zealand. We provided such countries with railways and many other things. If and when this equipment was paid for in goods and services it was barter trade, and there is much to be said in its favour. But in other cases the transactions were of a different character. They were controlled by finance. These transactions were somewhat complicated in form, but the final result was this. The equipment was paid for, not in goods, but in bonds bearing interest. Our financiers thus equipped and financed the general development of the countries, providing factories and workshops, competing and otherwise. They appear also in some cases to have provided for agricultural development: it is not only the Jews of Dróhicn that own herds, but the financial interests of Britain may control them in the Argentine and other countries. The process of equipment is at first sight admirable, and within certain limitations may be necessary and of value, though it would, I suggest, have been more valuable if the working producers of this country

and not the financiers had received the interest-bearing
bonds. In any case, we have to study its effects philo-
sophically, and not lose our heads attacking persons or
even systems.

Equipment trade has been of great importance in the
past, but it is a dwindling element in our trade, and so
far as it is going on to-day appears to be mainly concerned
in fitting up factories in our colonies to produce goods
that we have in the past provided from our own workshops. In this connection it may be noted that, although
there is considerable support in Australia for development
of trade within the limits of the Empire, the fundamental
policy of that country to-day is, and will be directed to,
making that country as self-supporting as possible. Only
with that end in view, and so far as it does not interfere
with the development of their country, they will willingly
trade with us. Indeed, one of the first steps taken by
Mr. Scullin on his becoming Labour Prime Minister of
Australia in 1930 was to invite British manufacturers to
set up their factories in that country. Accordingly, we
found in the following year that English firms were
spending large sums in enlarging their Australian factories
and building new ones, and there is every reason to believe
that they will continue to do so. (27) No doubt British
machinery-makers have equipped, and will in the future
equip, these factories, which may also be financed from
this country. A similar policy appears to have support
in Canada. Woollen factories have been recently set up,
and in 1931 more were contemplated (28), and it is at
least to be hoped that the equipment orders come to this
country. In South Africa, British manufacturers are

helping to create and equip coal and iron mines and factories with all that is necessary to provide that colony's needs for iron and steel. Equipment trade gave in the past, and will give in the present, work to Britain; it keeps up our exports and provides employment. But the ultimate result has been, or will be, disaster. For not only, so far as it helps the building up of competing industries, does it tend to destroy our markets, but out of the annual payments to this country of interest and dividends on the foreign investments arises a financial trade tribute, the economic effects of which are similar to the tributes of corn that long ago went far to destroy the agriculture of Italy and, finally, took its part in the decay of Rome. It is this form of trade that has now to be considered under the title of "Tribute Trade."

(c) TRIBUTE TRADE

We have now to consider what happens to the interest and dividends on our foreign investments. Such investments may be the outcome of equipment trade or other financial transactions, such as financing wars between nations (an example of the latter occurred in the war between Russia and Japan, when I understand British finance helped to provide funds for both sides). These payments appear to come to this country in some form of banker's drafts. The bankers distribute the amounts so received to the holders of investments for whom they are in the position of trustees; but they have to cash their own drafts. They have, dealing no doubt concurrently with certain expenses, such as insurance and bankers' charges, to get something in return from the country from which the drafts came. What they do finally is either to reinvest the money in the debtor country or buy directly or indirectly something produced by the peoples of the country concerned. "We generally cash such drafts in food," said a banker to me one day.

When we get to the bottom of all these complications we find that, whilst the capital remains abroad until it may be found convenient for it to be repaid, which may be at any moment,* a large amount of food and other goods come into this country to pay the interest and dividends on foreign investments. (29) This is what is called "Tribute Trade." We do not know the full details of these transactions; it is, of course, only the bankers who know this, but the result is that the people of perhaps a score of countries are paying us this tribute in the form of importations. Such importations, so far as they are in food and goods that we might produce in this country, cause widespread unemployment; this is especially true of agriculture, where importation of food may well have put a million workers out of employ. Thus, though there is probably some resulting employment in luxury and other trades in this country, important industries are undermined. Moreover, this form of trade may well be disastrous to the peoples of the exporting countries, who have to work hard to produce goods, not for their own consumption, but to go abroad.

To me, the most interesting, as it is the most obscure

* When this repayment comes and our foreign investments are liquidated, they will presumably be liquidated in the form of food and other goods.

of these combined financial and "tribute trade" transactions, is that of the trade with Denmark, on which
something may be said. The story of the reconstruction
and equipment of Denmark in the last half-century is
told by sentimental writers as a marvellous story of
Danish energy and capacity. But who, I have often
asked, financed it? I do not know: but the only person
I ever met who, owing to his knowledge of European
finance, might have found out for me thought it was
English capital, provided through Hamburg, but he
could not or would not tell me more. There seems no
doubt, however, that the reconstruction of Denmark
depended to some extent on foreign capital and cheap
semi-civilised gangs of labourers drawn from the Slav
populations of Central Europe. I do not complain of this
form of reconstruction. It seems at first to benefit everyone; the Slavs probably enjoyed their work in Denmark.
They came over in gangs, camped in farmers' barns, and
were fed, no doubt better than they would have been in
their own country; on Sundays they dressed up in their
peasant finery and all trooped off to a Catholic church,
if they could find one. Lutheran Danes, in their "Sunday
blacks," watched them with interest and amazement.
They danced and sang and fought amongst themselves;
and if now and again, as I am told was the case, a Slavonic
worker ran amuck and killed a Danish employer in some
obscure quarrel over money, he would be consigned to
a Danish prison, where life would be far easier than it
would have been if he had fallen into the hands of the
authorities of his own country. (30) In due time the
workers went home with money in their pockets. Now

what is the position to-day? The Danes are sending their food from their reconstructed country to England. It comes mainly in the form of bacon, butter and eggs. The peasants work hard nowadays, for prices are relatively low, whilst owing to the fear of unemployment very little Slavonic labour is allowed into the country. Wage rates seem lower than in Britain. The Danes themselves have only a second claim on their admirable produce; for, according to reports, they have been accustomed to consume the bacon condemned by government tests as unfit for exportation, to eat margarine instead of butter and to import eggs, probably of inferior quality, from other countries. I once asked a young Englishman living in Denmark if there were any new features in his district. "Yes," he said, "one of the peasant women has been eating butter; the whole countryside has been talking about it. Moreover," he added, "I saw six peasants divide an egg between them." Before the war eggs were imported from Siberia and a few years ago there was discussion in Denmark on the problem of Rumanian imports of eggs.

When I was in Denmark in 1929 I asked an intelligent peasant what he thought of it all. "It is a hard life," he said, "but I don't complain; there is the fresh air and the work." In another part of Denmark in the course of the same visit I stayed with an intelligent farmer who took me all round his district. He had travelled much and he owned a large and beautiful farmhouse, yet his children worked in the roughest clothes, with bare feet thrust into wooden shoes. He had, he said, made no profit on his pigs for two years, but he hung on. His wife filled the house up with paying guests; perhaps that is where the money came from. The latest reports from Denmark tell of increasing difficulties and poverty amongst the workers.

Only in the sugar-beet areas, it appeared, at that time were the Danes allowed to use Slavonic labour, which came from the country that is now Poland, and even that was rationed by the government. The percentages of Poles allowed, I was told, varied from time to time; it depended on the price, fixed to undercut if possible the foreign price. "If you want to compete with us in the sugar market," remarked a Danish friend to me, "you've got to follow the example of Denmark." I did not tumble to his remark at once. "Co-operation?" I hazarded. He laughed: "Oh no, not co-operation—cheap Polish labour."

Whilst the Danes themselves suffer, the result of Danish importation into Britain is also disastrous to this country. A business committee sits in London, week by week, and fixes the price of Danish bacon apparently so as to undercut the British farmer and to secure that at any rate he never gets a firm hold of the home market. This committee works in, it is said, with the multiple shops, and if the British farmers were to make a sustained effort to secure the market, the Danes would undoubtedly cut their prices; there would then be a long trade war in which everyone would suffer, but ultimately the Danes, with their lower standard of life and the command of Polish labour, would win. (31)

Whilst undoubtedly the Danish farmers are suffering to-day, Danish importations put at least 100,000 and possibly far more workers out of employ in the agriculture of this country and we have to pay the cost. Neither is there any considerable employment in British manufacturing districts to supply Denmark with what it needs. For if we look at the returns for the year 1930, the last year for which figures are available, we see that in that year £52,500,000 worth of goods and produce came into England from Denmark and £13,880,000 worth of coal and manufactured goods went back in return. The balance of over £38,000,000 is represented by food imports to Britain.

The problem of Danish trade with this country remains unexplored. How does it come about that the Danes send us over £52,000,000 worth of food and we only send under £14,000,000 worth of coal and manufactured goods in return? It may be there is a return of coal and goods through other countries; it may be that the trade is in part "tribute trade" arising from earlier investments in Denmark or other countries. But there may be yet another explanation of part of the transaction. The imports from Germany to Denmark in the same year were valued at £32,800,000 and the exports from Denmark to Germany were only £14,500,000. Now the Germans had to pay us in that year large sums in reparations: it may be that it was in the course of this transaction they exported goods to Denmark; then the Danes sent us food, and our payments to Denmark, passed on in part to Germany, were set off against the German account for reparations. We do not know, but, whatever the explanation, it is hard for free traders to argue that our trade with Denmark is an exchange of goods and services beneficial to both countries. Still they do. In the future they say it will be all right. Unfortunately for their argument we have to live in the present.

All the various forms of overseas trade described above together with certain others, such as, for example, trade created by repayment of investments, of which more may be heard in a few years, (32) go on concurrently with very varied results, some beneficial to the productive workers and some so disastrous, as in the case of tribute trade reaching this country in the form of food, as to put hundreds of thousands out of employ.

These overseas trade transactions, of whatever nature, have to secure a certain measure of financial support or they could not be carried on. In the background, then, are the financiers, national and international, with their varied interests, exercising influence and in some cases control in many different ways. Their power, such as it is, is no doubt generally exercised in the interests of the traders, but not necessarily in those of the manufacturers or the workers concerned. Financiers may, indeed, refuse support to transactions that would be beneficial to producers and consumers, and give their support to others that are disastrous in their ultimate results. They may also, by raising or lowering rates of interest, by causing or allowing alterations in the value of money or by transfer of gold reserves from one country to another, help or hamper trade transactions.

There is to-day much talk of "money power" as the all-controlling influence in trade and industry. It is true that finance exercises a control over trade and industry.

But so far as it is a control, it is not always intelligent and responsible control, and the results are often obviously entirely different from what is intended or might be expected. There are innumerable conflicting financial interests, often counteracting one another; and these in their turn are often influenced by political considerations.

The subject is dealt with later in this essay, but a metaphor, slightly fantastic, may help the imagination. Trade may be compared to a somewhat ill-equipped sailing ship on a somewhat stormy sea, working its way to a definite port, but yet, on its passage, driven this way and that by fitful gusts of wind, freshening now and then to a gale, the breath in its sails of the Æolus of finance.

The general result of the conflict of financial interests that thus arises is chaos, which on occasions creates masses of unemployment: it is not therefore surprising that governments intervene on the one hand with bonuses, to endeavour to encourage exportation of surplus goods, and on the other hand with tariffs, or regulation or prohibition of importations, to endeavour to prevent such forms of trade as appear to be creating unemployment in their own countries and so to undermine their civilisation. Alone of all countries, Britain, obsessed by the strange idea that trade of necessity created national wealth, did but little to safeguard herself against the disasters that arose from unregulated overseas trade. Thus much of overseas trade.

3. HOME TRADE

The internal trade of our country could be analysed in the same way as the overseas trade, for the home and overseas trades are really only two branches of the same tree; in both cases trade consists normally of transactions between individuals or groups of individuals of which the most important forms can be analysed as barter, as equipment and as tribute trade.

But nothing will be gained by going over the same ground again. Moreover, most people look upon home trade in a different way from that in which they think of overseas trade, and I propose, since imagination only works easily in customary channels, to fall into line with customary thought and discuss home trade from what I believe to be the point of view of the public. It will, I think, help to make my argument clearer if we go back to the ideas I have dealt with in the introductory part of this book and then carry on the story in its relation to home trade to-day.

When free trade first began to be considered in this country in the XVIIIth century, the movement started with an attack by the economists of the new school of the time on our traditional national policy of trade regulation, with its two aspects of price-fixing and control of distribution. It had in the past been a common, though not, of course, universal practice to fix fair prices as between producers and consumers by independent committees or authorities. Further, so far as corn was concerned, this internal price-fixing was in the XVIIIth century supported by methods of regulating imports and exports by duties so as to secure so far as possible that when there was a shortage of corn in this country importations would be encouraged, and when, on the other hand, there was a surplus, this surplus could be exported. (33)

There were occasions in the past when a government, during times of severe shortage, would itself buy corn. The opponents of free trade were then, as they are now, specifically in favour of a policy of controlling prices for corn and bread, and of so regulating the corn trade as to give the home farmer a first claim on the home market; the free traders at that time appear to have thought that prices, if left alone, would settle themselves automatically at a fair figure.

The second part of our traditional national policy dealt with distribution and formed part of the common law of England. Although the old industrial guilds had dealt with the distribution problem in detail in their rules, there was no definite comprehensive scheme of scientific organisation of distribution such as may be built up in Britain in the coal trade, or such as is also under consideration for agriculture in this and other countries. Nevertheless, the law was still quite clear that in dealing with food supplies, at any rate, no unnecessary dealers and merchants were to be permitted to come between producer and consumer. Such people were declared to be "oppressors of the poor and enemies of the country," and were to be put into the pillory, or fined and imprisoned as the law directed. Finally, it will be remembered, the controversy on these points centred round the trade in food, particularly corn. If you dealt in corn you could not make a profit; that was clear in law. Even as late as 1796 a man named Battam was tried at Aylesbury for making a profit of 17s. 6d. on a deal in about forty quarters of corn and was fined £200, a large sum in those days, far above what it is in present money values; into the bargain, he was sent to jail and told that he could stay there till he paid. The crime of dealing in food, explained the Recorder of Dublin in his address to the jury some four years later, was "not one which has for its object the injury of a single individual or even of a single class or description of person in the community; it reaches every order of the state and embraces at once the rich and the poor, the humble and the exalted." (34) The result of such regulation in the wheat trade was that the farmer could sell to the miller at a price deemed to be fair, the miller could sell flour to the baker at a fair price and the baker could sell to the public. The baker could not even sell surplus flour to a fellow baker at a profit. That was the law; but at the same time, no doubt, the practice was breaking down. Moreover, the supporters of this regulative system were not in a strong position since they did not know what was really the best machinery for securing its success. The free traders with the support of the economists, on the other hand, were against anything that regulated prices and distribution. They reiterated their view that if prices and distribution were left alone, prices would automatically adjust themselves to a fair level and free competition amongst dealers and middlemen, whose real business, they argued, was to bring producer and consumer together, would result in a thoroughly efficient orderly system of distribution at the minimum cost. Then all would be well. The results turned out to be almost the exact opposite of the predictions. For the new system resulted in a general chaos, a perpetual oscillation of prices and a great increase in the cost of distribution, and all was not well.

I propose to show by actual examples how competitive trade failed to secure what its promoters promised— what it means, not in theory, but in fact. Thereafter it will appear that it did not create the wealth of nations, nor even the wealth of Britain: rather that it created wealth for financiers and traders.

We are told by the orthodox economists of the old school that middlemen perform an important national service; certainly they do. Moreover, they are in a large number of cases men of special intelligence and capacity. I always defend them, for I know from my own experience as a middleman everything there is to be said in their favour. Of course, there are black sheep amongst them, but there are black sheep everywhere. I think, nevertheless, that looking at the position of middlemen from the point of view of national interests there is something wrong about their outlook and their system. I know that when I have been in the course of business engaged in a deal, the object at the back of my mind was always the same—to buy something at less that its value, and to sell it for more; and I confess that I was generally gleeful when I succeeded. But, to tell the truth, I was never very clever about it. I had been brought up with a highly developed Nonconformist conscience which, on occasions, troubled me and sometimes spoilt the pleasure of the deal. But such a conscience seems to me rare. Indeed, I have seldom met anyone in life who was not delighted when he had made a good bargain, and we all tell one another stories of our glorious successes. We are natural gamblers and most of us revel in it.

Recognising, then, that dealers and middlemen are not

criminals, we will turn to consider internal trade as carried on under the modern system in practice. We shall see whether it produced the perfect system that had been predicted, and really helped to create the wealth of Britain.

When I was in business in London in the early days of the present century, I was engaged in creating a centre to sell the work of artists and craftsmen in order to carry on the movement for the revival of the crafts created by the artist-craftsman William Morris and the group associated with him. At the same time I did a certain amount of miscellaneous dealing in all sorts of works of art and some furnishing. I was concerned in making enough money out of these side-lines to pay for the losses in the Arts and Crafts branch. Shopkeepers with taste, from Regent Street, Oxford Street, Bond Street and Piccadilly, and other fashionable streets, would come to me to purchase more beautiful things than they could provide themselves. As one of them remarked, "I sell ugly things myself in order to buy beautiful things from you." My business depended largely on making my little gallery attractive, and that in its turn depended on my being able to say smart things and tell amusing stories. People came for the talk and to see what I had to show them. I hoped they would remain to buy. Sometimes they did. Talk was my method of advertisement; it cost nothing. My friends in the trade, whom I managed to amuse, were very good-natured and gave me much advice. Thus I learned a good deal about retail trade in London. "You cannot sell this sort of thing in London," my shopkeeping friends would say to me, looking round my little gallery, "unless you charge at least twice what you pay for them." Then they would give me tips which were often very useful. For example, I remember using in the furnishing branch of my business a certain material for curtains. The first time I bought this material I went innocently into a well-known shop and purchased a few yards at 6s. a yard. Then some friendly customer, himself in the furnishing business, suggested to me that if I went round to the back door I could get it in longer lengths at 4s. a yard. This I did, but my friend said to me one day: "Why buy at the shop? It is not their material, it is only sold in their name. Go across to the back street, and almost exactly opposite the back door of the shop you will find a factor. You can buy the material much cheaper there." I did as he said. If I could have found the manufacturers the price of the material would, no doubt, have been even less. I remembered from my early studies of economics that the function of the middleman was to bring the producer and the consumer together, but when I tried to find the actual manufacturer of materials in which I dealt, I rarely succeeded; the middleman was the barrier. And when in my innocence I really brought producer and consumer together and introduced a customer to the artist-craftsman whose work I sold, I found my customer drifted off to the craftsman and I saw him or her no more. Then I sympathised with my fellow middlemen, who would not let me know where the goods they provided were made. My enterprise really failed because I could not make myself into an effective barrier between producer and consumer like the model middleman of to-day. I was, in fact, the ideal middleman of the economic theory. I brought producer and consumer together, and so my enterprise failed.

There is nothing immoral in the modern middleman's system; it is the natural outcome of adopting the free trade idea, but its effect on the price system is curious. There were so many prices and so many possible commissions. I soon learned to have a variety of prices myself, according as I sold to a customer, to a trader, or for export. I sometimes got quaint experiences—curious illustrations of the system. I once took a client and friend of mine into a well-known shop—I will not say where. I gave my card as I went in. My friend spent a little over £200; he gave a cheque, and as soon as it was cashed the shop sent me their cheque for £40, my share of the spoil. I had made no arrangement for a commission, and had done nothing whatever, unless one counts the couple of hours I spent very pleasantly talking about music and art, which in those days were my primary interest. Nevertheless I was essentially what in humbler circles is called a "tout." All the middlemen were in the same boat and were quite pleased to help one another. We were not the villains of melodrama that middlemen are said to be—nor are the financiers, as is sometimes suggested, the real villains of the play. We had to make our own livings; most of us had families to support—so, indeed, have the financiers. But there was something wrong about the system from the point of view of national service.

I was always being told amusing stories. A lady had brought a pair of shoes for repairs into the shop of a little London shoemaker I knew very well, for he was a countryman from my own county. She had complained that they had worn very badly; this annoyed her, as she had bought them at a fashionable shop and paid 50s. for them. The shoemaker, showing me the shoes, remarked: "They happen to have the maker's mark in them," and then he added, "The wholesale price is 10s."

A good-looking old workman who used to do odd jobs for me, a great gossip, remarked to me one day, "When I was young I thought I would not be a worker, so I tried for another job. I was a well-set-up young fellow and I got a job in a fashionable ladies' shop, opening doors to customers and putting ladies into their carriages when they went out. A lady came in for a pair of stays, and as she could not find what she wanted, the shopman promised to get them for her. I put her into her carriage and was then sent round to Soho and bought a pair from a French maker; the invoice, which I saw, was 3s. The shopman put some ribbon round the stays and sent me off to Grosvenor Square. I was kept waiting a little while, and then a cheque came down: it was for £3 3s." Possibly that was an extreme case. Here is a story of a later date from a big importer. He had a large consignment from, I believe, Central Europe, of very beautiful ladies' garments—dresses, then fashionable, made in one piece. He sold two consignments, each of a thousand, to two big London shops at £500 each for the lot—10s. for each garment. One shop had a special sale and spent hundreds in advertising and cleared them all out promptly; the price, so the advertisements said, was £2 10s. each. The other shop, rather more fashionable and distinguished, sold them as something very special at £2 each. My

informant thought the net result would be about the same
in each case. The importer was a clever business man
and must have made his own profit; one wonders what
the actual manufacturers got: probably about 10 per cent.
of the consumer's price.

One could tell similar stories of English manufactured
goods, but it would not be discreet. For the moment, it
maybe, the trade chaos has undermined the system, but
it will revive.

Perhaps these things do not seem to matter very much
since they refer to the luxury trades. But they do have
an effect in keeping down the wages of the manual workers,
and by reducing their purchasing power, dislocate the
relation between production and consumption, which, as
will be explained later, is a cause of unemployment.

We will now leave the luxury trades and take the home
trade in food, a basic necessity, where the effect of unregulated trade is far more serious.

This branch of home trade is, of course, kept in complete
disorder by the fact that, under our system of free imports,
all sorts of food are being sent into this country, either
to pay the interest and dividends on foreign investments,
or perhaps for reparations, or else for political or other
reasons, with or without special bounties from the
government of the exporting country. In many cases the
importers and their financiers and the corresponding
shipping agencies are, automatically, though perhaps
unconsciously, concerned in undermining the British
agriculturists so as to secure the market for the foreigner.
In the face of this attack our national trade policy gives
to the overseas trade the first claim on the home market,

and leaves the British farmer to look after himself. To deal with this situation the farmer is powerless; he is like a rat in a trap. Nevertheless, such is the indomitable energy and ability of our people that farmers do struggle on. Doubtless they would not do so if they could get out, but a man trained as a farmer has very little chance of starting afresh in other trades. It is important, in considering the food trade, to have all this in mind.

We will now turn to details. Taking the wheat and bread trade first, it is interesting to note that one effect of free trade has been the closing down of country mills (35) and the setting up of many important milling businesses on the coastline. Mills there instituted can get their wheat from abroad and also send the so-called "millers' offals," an invaluable food for fowls and pigs, to foreign countries by ship. Thus they leave the British farmer in the lurch and at the same time throw upon the nation the costs of the resulting unemployment. This development appears, in conjunction with the importation of flour, to have resulted in there being too many mills in this country; consequently, notwithstanding the closing of the country mills, many of the mills that remain appear to be working below their full capacity and so at unnecessarily high costs, reflected in the price of every sack of flour that goes to the bakers. This high cost is increased by the extreme uncertainty of prices, for wheat prices vary from day to day (36) and these variations have to be ensured against. The same uncertainty controls the baking trade, with a result that millers have, I am told, many bad debts, which have also to be allowed for. Amongst bakers there are similar difficulties, including, in working-class neighbourhoods at any rate, and in times of depression, a profusion of long accounts and bad debts; these difficulties are again reflected in higher prices. Here also, under our system of unregulated trade, there is excessive competition. The old-fashioned economists used to tell us that competition lowered prices. Sometimes it does, but more usually in the retail trade it raises prices. I remember that, when I was speaking round England on these subjects some few years ago, I had two illuminating conversations with individuals in the baking trade. One was with an intelligent woman whom I came across in a little village meeting. I always encouraged everyone to talk at meetings that I held, and in this case someone in the course of the talk began attacking bakers. When they had finished, the young woman remarked quietly, "You need not attack the bakers." I asked her to explain, and after a little persuasion she did so. "I am the daughter of a baker in the East of London," she told the villagers. "I am here on a visit. We had," she went on, "a small area which we divided with another baker; we were quite friendly; we both charged 9½d. for the four-pound loaf"—a fair price in those days. "Then a third baker came in and cut the price to 9d. We in our business had the same general expenses, but a smaller trade. All the three bakers suffered together, but for three months we went on; then we all raised the price to 10d.—a halfpenny a loaf more than it had been before the new baker came in." Competitive trade in bread had not lowered the price, but raised it. The other story concerns a country baker. I rather think he was a Methodist. He did not talk in the meeting, he was too well known and did not wish to give away his affairs; but he talked to me afterwards. His customers, I gathered, were mostly chapel people, scattered over a large district. His cart went from village to village, wherever there were such people, but he had other scattered customers, and he described what this widespread distribution meant to him in costs. Finally he said: "If I had just a village or two near my bakery to supply, as I had during the War, I could reduce the price of bread and even then do better than I do now." He also was the victim of competitive trade. Moreover, he had to safeguard himself against other bakers cutting in; the trade war was always at the back of his mind and he charged accordingly. Even then he was not making his fortune.

What is the result to the public of this general confusion? In the autumn of 1931, when wheat was under 25s. a quarter, there was, after allowing for the offals sold off, less than twopennyworth of wheat in a four-pound loaf. According to Sir Charles Fielding, (37) if milling and baking were put on a sound business system and the main evils of competition eliminated, the whole cost of converting wheat into bread, including a fair margin of profit, should not be more than 2d. Mr. A. H. Hurst, in his important work on *The Bread of Britain*, (38) points out that the combined charges for milling and baking had increased by approximately 3d. as between the years 1914 and 1929, during which period "the unit cost of manufacture throughout the world has been greatly reduced." Incidentally, he goes some way to confirm the accuracy of Sir Charles Fielding's figure. If these analyses are correct, it follows that if our systems of milling and baking and relative distribution were perfected and organised, the price of the four-pound loaf at that time might have been 4d. for cash over the counter. If, however, we make the more generous allowance of 3d. for the cost of milling and baking, the price would be 5d. over the counter, or 5½d. delivered. Nevertheless, in 1931, when wheat went down to 20s. a quarter, the actual price for bread delivered was 6½d. It is difficult to say how much a difference of 1d. in the price would cost to the consumers of bread; perhaps £10,000,000 a year. (39) High costs of milling and baking are not peculiar to this country. Gide, the French economist, suggested in the course of a discussion of the food problem, (40) that competition amongst bakers in Paris had resulted in a rise of the price of bread of 40 per cent.—a rise even higher than that observed in this country.

There is no reason to suppose that millers and bakers benefit from this absence of perfected organisation that produces the high prices. Bakers certainly do not—we do not find bakers amongst the millionaires. The high cost of bread, if we are to accept Mr. Hurst's analysis, is not due to the cost of wheat or to undue profits in milling or baking: it is due to imperfect arrangements, it is the cost of chaos. But do not, therefore, assume that it is advantageous to the nation to reduce the price of bread lower than 6½d. a four-pound loaf. The workers are not likely to benefit by that. Very low prices are, it will be seen, a disaster to the nation, and in fact, when prices of the essential commodities go down below a fair figure, wages go down with them. What is wanted is to keep prices at about 6½d. or 7d. a four-pound loaf and to raise the farmer's share of the price. The general effect of want of correlation in the trade in wheat, flour and bread is that the farmers get too little and the housewives have as a general rule in the past, though not always, paid too much.

I do not propose to deal further with this special question, but will explain and illustrate some of the effects of competitive trade on the business side of agriculture, with special relation to the farmer's life and work. Wheat may be referred to first.

It will be remembered that Cobden, himself a farmer's son, explained to the country people that free trade would bring about the stabilisation of prices; the price of wheat on which he based his case would, he predicted, be about 45s. a quarter. But the year after the passing of the Act of Parliament for the introduction of free trade in corn, the price, as has already been noted, went up to over 100s. a quarter, with bread 1s. a quartern loaf; thereafter the price of wheat ran down to about a third of that amount, only to go up again to over 80s., whilst before half a century had passed it had come down to as low as 17s. 6d. a quarter. Moreover, during all that time it varied every day, and indeed every hour and in every market. So much for Cobden's promise of wheat prices stabilised by free trade at a fair rate.

Of course, this complete failure of free trade policy, so far as it affected wheat prices, made no difference to the speeches of the free traders, for by that time belief in free trade was an "act of faith."

A variation of prices, similar in character to that

affecting wheat, is universal in agriculture, and it is easy to analyse the result.

Farming has long since ceased to be a business proposition; it is a gamble. Indeed, the most exact parallel in economics to farming is horse racing. The farmer is the backer; if in the gamble of the market he gets his crop to market at the moment that the price is high, he has backed a winner and he comes home with his winnings in his pocket. But, as in horse racing, the backers seldom win unless they have inside information, so in farming the farmer rarely has the luck and makes his haul; on the whole he loses, and if he loses—well, he has lost money over his wheat, and then, with indomitable energy, he tries something else. He turns his land down to rough grazing, dismisses his men and limits his production to a minimum; creates unemployment and reduces national wealth. The wealth of Britain suffers but he cannot help that. Then perhaps he does better: he may back a winner this time, but other farmers have taken the same line, and again he may find that he has backed a loser; this time he may be ruined. As an alternative, if he is clever enough, he may take up dealing and become a "farmer dealer." In that case, when the slumps come, he may lose as a farmer but flourish as a dealer, for in such times clever dealers flourish; they can buy cheap from half-ruined farmers. They are not only within their right, but are especially protected by various old Acts of Parliament, which give legal status to the dealers.

The prices of wheat and bread, on which so much of the life of the nation depends, fluctuate in the main with variations of supply complicated by general muddle. We can take the potato trade to illustrate further the effect on prices of variations of supply. Potatoes varied in price in the decade 1920–30 between about £14 and 17s. a ton; the last price is really a nominal quotation, for at times the bottom dropped out of the market altogether, and then potatoes could not be sold at any price. Of course, it is all right for the farmer when the price is high. There used to be, and perhaps still is, a certain smallholder in the fen country who for many years, by a strange accident of fate, always had the luck of the market, and confessed to having made £15,000 in a very few years. He happened to have always backed the right horse and won. Talking of the potato trade, he told how one year, not so very long ago, he had had twenty acres of potatoes, got ten tons per acre and sold them for £10 per ton. He got £2,000 for that one crop. He certainly pocketed £1,000 profit—probably a good deal more. When I was talking of this deal a year or two later to a neighbouring farmer, he remarked, "I sold my last crop of potatoes for £1 per ton, and thought myself lucky to get it."

When the bottom dropped out of the market some years ago, a well-known Kent potato-grower decided not to dig his potatoes at all, and told the people of Orpington that they could take them away; they came out, I am told, in crowds and grubbed them up; they at least benefited. Coming up to London from Kent at that very time, I asked a Hampstead housewife what she was paying for potatoes. "Potatoes are very cheap," she said, "only a shilling a stone for the best"—that is to say, £8 per ton. What did the greengrocer say, in friendly conversation? "Of course," was his explanation, "I can get potatoes for next to nothing, but what am I to do? If I put my potatoes down to $\frac{1}{2}$d. a pound, the whole trade would be upset; and though my customers might be pleased to-day, they would think there was some fraud when I had to put them up again to 1d. or perhaps later to 1$\frac{1}{2}$d. or even 2d. In this business," he continued, "I must make a profit when I can; it's none too easy—extremely speculative. If I had a fair, steady wholesale price I could keep my retail price where it is and everyone would be satisfied."

In almost all market-gardening crops the position is very much the same as that which I have described in relation to potatoes, though, of course, exceptionally clever men, by direct delivery and other special methods, find their own personal solution of their individual problems. Sometimes the trouble arises from irregularity of supply and demand, sometimes from what is called "bargaining power" and sometimes from other causes, but all the difficulties have their origin in the adoption in this country of the policy which leaves the trade entirely unregulated.

We can consider a few examples of the difficulties that arise in the fruit and vegetable trade, illustrating them by stories told by my friends at Covent Garden and elsewhere. Cabbages, I am told, are sold at Covent Garden and other markets in queer sacks called "Matts," with one side longer than the other, on to which the cabbages are rolled out for inspection. "There was a glut of 'Matts' in the market this morning," said a friend of mine in the trade, who was coming back one morning some years ago ago to a late breakfast, just as I was going off to work. "Our firm bought all they could deal with at a low price;

the balance, I suppose, went to costermongers or was thrown away." "What will happen next?" I asked. He explained. On the morrow, he expected, it would be just the same; then the market-gardeners would "get the wind up"; they would send no more cabbages to Covent Garden and in a day or two the market would be empty. Cabbages would go up to fancy prices; then this bit of news would in its turn get round the countryside, and as a result every cabbage-grower within twenty miles of London would cut his cabbages and shoot them into the market; it would be swamped again. "It is a rotten business," said he; "it's not only the market-gardeners who suffer—it's bad for everyone, ourselves included."

Sometimes even stranger situations arise. A friend of mine, well known in the flower market, who was full of genial gossip, told me—it was before the War—this very illuminating story. It appeared that very late one night the news got round in Covent Garden market that there was a likelihood of there being a glut in the tomato market next morning. Now at that time there were two groups of merchants in this branch of the trade who more or less controlled it; a buying and a selling group. The two groups collected at two separate hotels, and messages went backwards and forwards all night long trying to agree to terms. No agreement was arrived at, and when the tomato market opened in the early morning the selling group withdrew all the tomatoes. What they did with them my friend did not know. But the quarrel, he said, went on all the week, with disastrous results to both producers and consumers. The former got nothing, the latter had to pay higher prices or do without. To the economist who says that prices are settled by supply and demand, one can tell this story as an example of how the normal effect of supply and demand is defeated by what has been called "Bargaining Power."

It is, of course, a commonplace that in the course of this entirely unregulated trade the market-gardener who sends up his cabbages or tomatoes, his marrows or his carrots, may get, on individual deals, nothing whatever, and sometimes even a bill for expenses; it is also a commonplace that prices may disappear altogether, and it may not pay the producer even to pick his fruit, or gather in his vegetables, which are left to rot.

Thus much for vegetables and fruit; we will now take up the pig trade, and again I propose to illustrate the general situation by minor incidents, at first sight trivial. I make no apology. These small matters are not to be ignored, for it is by the study of comparative trifles that you get a revelation of the facts from which general principles are evolved: indeed, such a study may tell you, if you have the power of analysis, far more than the most brilliant treatise on political economy. It is well to realise this and to remember that these treatises are often prepared by individuals who have no first-hand knowledge of the facts. I suspect, indeed, that many are in the position of the most learned and distinguished of my friends— a man whose words of wisdom have been followed and accepted without question by his many admirers. He is more frank than most men of his class, for he confessed in his old age that his great difficulty had always been that he had never been able to discover how trade and finance

were actually carried on. I always want to introduce these
great thinkers, men of delicacy and refinement and often
complete detachment from the facts, to my friend the
pig-dealer's tout. Probably, if I were so fortunate as to
be able to effect an introduction, the distinguished economist would look with disdain on the tout and consider him
as a low fellow. He would be wrong, for the pig-dealer's
tout is not only a thoroughly good fellow, but could
provide the distinguished economist with illuminating
evidence of the failure of the "law of supply and demand"
—a failure which, as a fact, occurs in all branches of
trade, and would cause him to recognise the importance,
in matters of price, of bargaining power. In any case,
we will make the acquaintance of my friend. He is to
be found at the "Jolly Cricketers," his house of call on
most Saturday nights. There you may also find his colleague, in a similar line of business—the book-maker's
tout. The former, to make an economic analysis, deals
with the necessities of life; the latter is in the luxury
trade. You can find men of similar occupation concerned
in bringing buyers and sellers together and aiding nego-
tiations throughout trade, and even meet them in social
life; mostly good talkers and many spotlessly dressed and
with charming manners. Both the pig-dealer's and book-
maker's touts are strangely typical of a system. They
stand for the real things of life.

My friend the pig-dealer's tout is not spotlessly dressed,
but he is a good talker and has a kind heart. One evening
he talked to me of "dealing as an art," though he did
not use that phrase, and described his day's work. It
appeared that in his district the dealers divided up an

area between them, so that if his employer did not succeed in making a bargain for the pigs he wanted to buy from a particular farmer, no other dealer would make an offer for them. The farmer would then have to take his pigs to market, with all the risks this involved. On the day when we met at the village inn, the pig-dealer and his tout had visited a certain farmer who had pigs to sell. The conversation in which the tout took an active part turned naturally on the low price of pigs. At last the dealer made an offer; it was not, I gathered, a high one, but the tout held up his hands in horror and told a story of pigs selling for next to nothing in a neighbouring market; and so the conversation went on. The farmer was busy, hard-pressed for money and ignorant of the market prices. At last the bargain was finished and the dealer got his pigs at well below a fair price. Here again is a humble illustration of the middleman's actual function, which has been already analysed. The pig-dealer may be hard-driven to make a living and know no other way to make an honest one—he probably had a wife and family and did not want to drift into the unemployed; there is no dole for pig-dealers; if he breaks he may never recover and may end in the workhouse. He looks on the farmer as fair game, as does the sportsman who is hunting the deer. He may indeed well have the same affection for, and interest in, the farmer as the sportsman has for the deer. One hears, in fact, many stories of the kindness and thoughtfulness of country dealers. The particular pig-dealer's tout, at any rate, was really sympathetic to the farmer, but he had taken up the job after a long spell of unemployment and could find no other work. This story gives a specific case of that widespread influence in price-fixing and economics—bargaining power.

Economists who want to learn something of the economics of facts should undertake an intensive study of the pig trade and see how it affects the wealth of nations. There are many things to investigate besides the doings of the dealers. First the complications that arise from the varying values of pork and bacon; then there is the committee sitting in London, representing Danish interests and fixing prices to suit their branch of the bacon trade; then there is the perpetual vacillation of prices. Pigs can be created very quickly—a sow may farrow fourteen months from her birth and may have two litters within twelve months. As a result, when prices go up, it used to be the rule, at any rate before the War, for all the little people in England, and some of the big ones too, to get one or more breeding sows; soon the villages would be full of sucking pigs. In East Anglia everyone on such occasions used to be smiling. "Pigs are up," was the cry. But peasants in other countries would want to take advantage of the higher price, and there would soon be, not only an increase of home supply of pigs, but also of foreign supply. Prices would then go down with a run, the breeding sows might go to the butcher, and for a time pig-breeding would be at a discount. A few people in England would have done well out of the gamble, others would have lost. Peasants in Europe may have suffered similarly. Finally, there would have been little relative gain to the consumer from the slump, for retail prices would not have gone down in proportion.

It is estimated that this muddle, reducing output, in

F

the pig trade keeps about 120,000 workers out of employ, in this country, in the actual business of pig production and in bacon factories and other relative trades. This perhaps costs the country £6,000,000 a year in unemployment benefit, whilst under-production may be estimated at £60,000,000 a year at wholesale prices. (41)

I will tell one more story of the countryside. Speaking once at a village meeting, I said something about finance and told the story of the herd of cows in Popina and the Jews of Dróhicyn. The parson of a neighbouring village capped the story. Somehow he seemed to have missed my point, for he told of a farmer in his parish and his flock of sheep. It appeared that a local banker had financed the flock. "How kind of the banker," said the parson, "to help the farmer in that way!" A young farmer whispered in my ear. "I always try to keep out of the hands of those people," he said crudely. It appears that much depends on how you look at these things. (41a)

Every day in the press examples are given of the confusion in the production and distribution of food products and of the disasters that occur: it is hardly necessary to stress the point. What learned people used to tell us was that these conditions were abnormal and did not occur in other trades. I am inclined to think that, though the position in agriculture is picturesque and fantastic to a degree that one does not find in other trades, the same evils may occur in other industries even on a larger scale.

I give two incidents to illustrate this. The first arises out of a law suit. A lady in Liverpool recently sued a firm who had supplied her with garments which, owing to some chemical ingredient in the colouring-matter with which they had been dyed, caused her to be afflicted with a skin disease. The firm counterclaimed against a factor who had supplied the goods. This factor claimed against one in London, he against another, and so it went on, until the ultimate claim was against a firm in Central Europe. A strange and illuminating story. The second story is out of my own experience. Last time I was in the coal-mining district—it was some years ago—I was staying with an out-of-work coal-miner, a man of intelligence and some education. He described the life of the pit—rather an old-fashioned one—his long walk underground, and then at last getting to work. He told me that, after allowing for various customary and other deductions from his wages, he was only getting on the average a net 5s. a ton for his output; the coal, I found out on enquiry, was quoted retail in the London market at 55s. a ton. No doubt in the coal trade, and in all the various industries that go to make up the clothing trade, which deserves more study than it has yet received from economists, distribution is more systematically organised than it is in the food trade, but even in these cases we have vacillation of prices, well illustrated in the case of the clothing trade by the sporadic "Sales" at reduced prices; in both the cases the machinery of distribution has certainly taken a form that is extraordinarily costly. I am not, indeed, at all sure that the share of the retail price of food that farmer and labourer divide between them for their productive services is not larger than what is secured by the productive worker in these other branches of trade. Probably in normal times farmer and labourer get for these services, on an average, somewhere about one-third of the retail price, and the other interests, representing distribution and conversion into food, with finance, rent, tithes, rates and taxes, take two-thirds. A great deal is said by land reformers of the claims of the landlord, but though rent and tithe may loom largely in the farmer's budget, their proportion of the retail price is probably little more than 5 per cent. (42)

I propose to tell a final story, not only for its own sake, but in order than my readers may see how the telling of illustrative stories creates in men's minds an entirely wrong point of view.

I was standing one day on the coast in the Ballyconneely peninsula, south-west of Clifden in Galway, watching the Atlantic rolling in, whilst a gang of sturdy Irish peasants, big black-headed men, at that time ardent followers of De Valera, were raking in kelp—seaweed, as we call it in this country; they burnt it in rough stone kilns and loaded the ashes up on carts. It was ultimately, I understood, to be converted into iodine. "And how much do you get for it?" I asked one of the men. "Sorr," he replied, with a rich brogue that readers must create for themselves, "it is this way: the dealers pay us £4 a ton; the manufacturers pay £32 a ton, and as to the public," he said with a fine sweep of his arm, "they pay a guinea an ounce." Such is the Irishman's story; there is at least a grain of truth in it. I read this story from my manuscript to an Englishman. "Organised robbery, I call it," was his remark; "there are too many middlemen everywhere." And then he dismissed it from his mind. But he was wrong in his analysis, or want of analysis, and he should not have dismissed it from his mind. Let us try and analyse. To begin with, the very word "organised" shirks the issue because the main point is that the trouble arises from want of organisation: it is really the outcome of disorganisation, unregulated distribution, competitive or free trade in its simplest form. Further, it is not "robbery," save in the sense that all property is robbery, and such theories do not help us to solve the practical problems of the day. Finally, it is not true that there are too many middlemen everywhere. There may have been in this particular transaction, and here and there we may find a surplus; but if we consider, for example, the various processes that lie between the home producer and the retailer of our food supplies, we shall find, if and when we come to reconstruct agriculture and create intermediate industries, such as bacon factories, we may want more intermediaries, not fewer, for there will be a greatly increased home trade to deal with. In any case, it would be dangerous in the national interest to base constructive policy on removal of middlemen; we should only create more unemployment.

Everyone has some knowledge of the cost and confusion that have spread through our distributive trades; most people, like the Englishman of my final story, are inclined, when specific facts come to their knowledge, to give little thought to the matter and to jump to conclusions: they generally close their discussion by an attack on middlemen —a popular stunt. Few seem to realise the need for very careful analysis of the facts. To secure the true lesson of these stories one has, then, to analyse and find out what is good in the transactions described and in the systems that lie behind them and where exactly the evil lies. I venture to suggest to such of my readers as have reached this point that they should go through these stories once more, and pick out those particular elements in the transactions that are valuable and lead to constructive proposals, and distinguish them from those that are evil and lead to chaos. Our main object is clearly to get trade out of its present highly artificial channels and get it back into natural channels, and so secure a real exchange of goods and services beneficial to all concerned. To my mind the intermediaries of all classes are the people to be consulted; they have full knowledge of the facts and are profoundly aware of the evils from which they themselves suffer. It is to them that the government should go for advice as to the building up of a better system. For such work of reconstruction the intermediaries will, I have no doubt, be found willing not only to advise but to help in carrying out constructive proposals; certainly they are profoundly concerned in securing a better system.

In the background of all the transactions described lies the question of motive; there is much to be learned in this relation from the story of two groups of dealers who on one occasion manipulated the tomato trade in Covent Garden market. The organisation of dealers into groups seems to have been in itself a good thing, but it was the motive that was wrong; had their motive been a different one, had they been concerned to carry through the transactions in the interest of producers and consumers, securing a fair price to both, all would have gone well. It is clearly not beyond the wit of man to secure such a change of motive. More will be said on how to secure our objective when constructive policy is dealt with in a later section of this book.

4. SOME GENERAL COMMENTS

Before concluding this section, some general comments may be made on the problems arising in the course of the discussion of "Trade in Theory and Fact."

A beginning was made by explaining that according to the orthodox theory trade was an exchange of goods and services, a thing beneficial in itself, and that being beneficial to mankind, it should be free. We find this analysis of trade to be a myth, for in fact trade is not a single thing, and the word covers a large number of different transactions, good, bad and indifferent; some producing wealth for the nation and employment for their people, others poverty, misery and unemployment. Trade is often not a constructor but a destroyer. We are perhaps also inclined to suspect that, under a system based on free competition, destructive forms of trade have developed rapidly and have brought disaster to the world. We may have observed also that many other popular theories appear to be untrue. Competition does not, as economists have said, necessarily lower prices; its effect varies, but one thing is certain, that it results in fluctuation of prices, whilst a not unusual effect of competition amongst traders, in the case of the necessities of life at any rate, is to push down prices against the producer and thrust them up against the consumer. We also realise, rather dimly perhaps, that prices are not fixed by supply and demand, as so many people assert; there are, in fact, many things that influence prices, but all we can see at present is that, whilst it is true that supply and demand influence prices, bargaining power also comes in. Moreover, there is an even more powerful cause of change of prices that has only been touched on—the variations in the value of money, which, by altering general price levels, produce perhaps the most startling results of all.

I sometimes wonder why it is that, although the axioms that the orthodox English economists have made popular are found to be untrue, we have up till almost the present day settled, and may indeed continue to base, our national policy upon conclusions themselves based on these untrue axioms. We have founded our policy on fallacies; we have built a house on the sand and then start buttressing it up: we have never looked at the foundations. We have been and indeed, at the time of writing, still seem to be anxious to develop our overseas trade, though it is easy to show by common sense or by reference to history that, with certain exceptions, the development of overseas trade to-day, even if it be possible, is almost certain, whatever its ephemeral results, to produce ultimately poverty, misery and unemployment on an even larger scale than it is to be found in our country to-day.

The explanation of this strange phenomenon of belief in what, if we think at all, we know to be untrue may be that we have not yet entirely escaped from the grip of the purely imaginative thought, amounting at times to hysteria, that has controlled our national policy for over a century. It is a strange thing to observe how long an entirely false idea continues to influence thought; there is something extraordinarily attractive about a fallacy—it fascinates like a bright musical comedy, or one of those obscure oriental religions that are fashionable in certain circles to-day—it takes one completely out of oneself and away from the hard realities of life. Truth can never really compete with it. It is easy for anyone to find examples of this power of fantastic ideas. I know a distinguished Quaker lady of brilliant intellectual powers whose life is inspired by the desire to bring the War of Arms to an end in this world, but is yet a confirmed believer in the Trade War. Bemused by the words "Free Trade," she allows herself to be controlled by the belief that there is some great moral impulse behind it. There are many such people who gather round Geneva and the League of Nations. Entirely illogical, they do not see that the War of Arms and the War of Trade are inspired by the same idea, and ultimately produce the same results.

Indeed, from one point of view, the effect of the Trade War is even worse than the effect of the War of Arms. There is certainly some element of heroism in the War of Arms, but there is none in the Trade War as it is carried on to-day: it destroys moral standards.

Finally, let us ask ourselves, in the simple Socratic method, a simple concrete question, limited in such a way that we can grasp its meaning—that is to say to our own country. What is the object of Britain's trade and industry? I suggest it is to secure wealth for Britain by production and distribution of the essentials and amenities of civilised life: homes, food, clothing, lighting and heating, health, knowledge and happiness. We have lost sight of that idea, and. bemused with vague ideas about the wealth of nations, have thought of trade as a thing, good in itself, to be developed for its own sake, when in fact it is often an evil. If Britain had, in the last hundred years, devoted half the energy that we gave to developing overseas trade to the work of securing these seven essentials of civilised life—good homes, food and clothing, a sufficiency of lighting and heating, and also a generous provision of health, knowledge and happiness for the people of our own country—we should have been the wealthiest and most highly civilised nation that the world has ever seen. And if we devote our energy to this object to-day, so great is the power in our hands that we may, in less than a generation, secure this position. We shall, no doubt, not then be pioneers in trade, but we may become, what is far more important, pioneers in civilisation.

PART II

SOME EFFECTS OF THE TRADE WAR

PART II

SOME EFFECTS OF THE TRADE WAR

I. INTRODUCTORY

Just as in the Great War the war spirit grew until it spread over the world and dominated the whole of civilisation, so in past history, when the competitive spirit in trade developed, it ultimately spread to every country. Competition is not necessarily an evil: under certain circumstances it is of value, but once a competitive struggle begins, although it may be obvious that if it exceeds reasonable limits it may well destroy civilisation, it seems almost impossible to stop it, or even to bring it under control.

We began the trade war gaily enough, the first in the field, a great adventure conducted in a crusading spirit: we went thoughtlessly on, wasting our natural resources, using up our coal, iron and other metals, and destroying our agricultural populations.

Our initial successes were remarkable, but a time came when opponents, often better equipped than ourselves, started against us. Then we lost our heads and allowed ourselves to be carried away by the war spirit: we forgot the objects of trade; we even lost the sense of a crusade; we were concerned to beat competitors; we talked of capturing markets; we thought in terms of the "knock-out blow." We fought, and at last we became so bemused that we ceased to fight for what we wanted—our homes and food and the necessities and amenities of life, in short the wealth of Britain; instead we fought for something that we did not want—overseas trade. It was the climax of this struggle when we endeavoured to arrange to send Canada coal we could not well spare, in order to get back corn that we could have produced ourselves. In this struggle we looked everywhere for new methods and new weapons, not to develop our own civilisation but to capture trade. Our manufacturers, with the backing of finance, turned their attention to every possible method of increasing output and reducing costs, in order to beat competitors, both at home and abroad. With this object they brought in science to invent labour-saving machinery and intelligence to reorganise production, and advertisement to persuade people to buy even what they did not want. Until quite recent years, business men pursued this method almost automatically, bemused by the idea that this development of production could go on indefinitely and that by some "law of nature" new markets would come out of space. And even when in 1929 unemployment was rapidly increasing and it was obvious that the limit of expansion of markets had been—for the time being at any rate—reached, the business men's only proposal was to restart the trade war, using the same old weapon, which they seemed to think would be effective since it had been cleverly christened with the new name of "Rationalisation." It was again the word that controlled thought: this was pure folly.

2. RATIONALISATION AND OVER-PRODUCTION

It is the policy of rationalisation that we have now to consider.

The effect of rationalisation, if it were not involved in our system of competitive trade, might well be to create wealth for the nation and leisure for the workers: two things, up to a point, admirable in themselves. But in fact under our present system, imbued as it is by the competitive spirit, rationalisation fails to increase to any great extent our national wealth, whilst it actually puts large numbers of workers out of employ.

Up to quite recent times the unemployment that arose from rationalisation (43) did not seriously disturb the minds of our industrial leaders. This complacence, no doubt, arose in part from the fact that the more influential leaders did not themselves suffer: they were like the Head Quarters Staff in an army, in the main protected. But it was also, no doubt, due in part to the belief of such leaders that unemployment, with all the loss and misery that it involved, was only a passing incident. If, indeed, the industrial leaders considered the problem seriously, they must have felt very much like a general who executed a manœuvre which lost his army a hundred thousand men in killed and wounded: to the general it would be an incident in the war which would be justified by ultimate events. In both cases it would have a different aspect to the sufferers, unemployed or killed, their wives and their children. To-day, fortunately, there is some change of thought, even amongst business men, which in due time will have an effect.

The nation, on the other hand, has been and is still to some extent befogged by the argument that the final effect of rationalisation will be reabsorption of the unemployed. This, in earlier days, had an appearance of truth; for so long as there were new countries to equip and new markets for export trade to be captured, the unemployed might be absorbed. Such absorption did in fact occur to a large extent in the XIXth century; but even then the re-employment was only of a temporary character, and was followed in a few years by more unemployment, which was partly absorbed by the distribution of population to our colonies. Economists, at a loss to find an explanation of these ups and downs (which they seemed to think was part of a law of nature), invented a phrase "trade cycles," (44) and left it at that. To-day the process of world development has either come to an end or is likely to go on so slowly that it does not really affect the situation, and any possibility of permanent increase of Britain's share in world trade is negligible. We have reached the saturation-point. With certain exceptions, other nations do not want our surplus goods, and it will not therefore help the situation if we rationalise our industries in order to produce them. Why should they want our goods? In many cases they can produce, or will shortly be able to produce, the greater part of their own needs: if they do not do this to their maximum capacity they are bound to have unemployment. Such countries are developing, and will continue to develop, their own industries; (45) whilst one special outcome of the new situation will be that more British industries will be moved to our dominions and colonies, where workshops and factories will be built up under the screen of tariffs or some other form of control against imports. All intelligent nations will encourage such development of home industries which not only produces employment, but helps to bring the national economic and financial problems under control—an essential element, as will be seen later in economic progress. (46) No amount of rationalisation on our part is likely to affect this general position or cause us to win in the competitive struggle.

There will, of course, always be some countries that will wish to buy manufactured goods from outside their own boundaries: in some cases they will accept British goods; this may be to the good in its increase of employment, but not of necessity, for they may send back in return agricultural produce or manufactured goods that we could produce ourselves, and so again create unemployment in this country. But it is not very likely that such purchasing countries will come to us, for there will be various centres of production to choose from. We might possibly, by drastic rationalisation, be able to make a fight in the markets of the world against Germany and the United States and other countries which have a standard of civilisation similar to our own; we might so secure some temporary increase of export trade. But it is not from such countries that competition will ultimately come. There are other nations with different standards. The possibilities of development in Russia create a special problem; but apart from Russia there are the Mid-European nations, Czecho-Slovakians and Poles, and other Slavonic races who may, for a generation at any rate, be quite content to work long hours at low wages. They will produce much more cheaply than we do, and yet, since many of them belong to a primitive civilisation with an undeveloped intellectual life, may get all they want out of life. Indeed, judging from my own experience of Slavonic peasants who seemed able to work with complete equanimity for incredibly long hours, such races may, thanks to their very want of civilisation, be happier, notwithstanding their long hours and low wages, than our own more civilised working classes, with their wider interests and far greater sensitiveness. The pleasures of such peoples—dancing, singing, music, love-making and drink—cost them, with the exception of the last item, little or nothing. No doubt the middle Europeans will take their share of such world trade as survives the next few years, but more may ultimately pass to countries where there is oriental and semi-oriental labour: to India, China and Japan. It is important, perhaps, to make this special tendency clear. It used to be said, and very many people still believe it, that the labour of skilled and intelligent Britons would always produce better results than could be obtained from cheap oriental or semi-oriental labour. (47) To a large extent this may have been true in the past, and it may still be so to some small extent in the future. But there is a new element in manufacture to be considered: modern machine processes are so perfected that the work of the individual is liable to become more and more mechanical; then, as it becomes more mechanical, it becomes more and more wearisome to intelligent workers, and more and more easy to the half-civilised Slav or the philosophic oriental. I have heard it said that some of the work in Ford's workshops in the United States is so mechanical that it can only be done by an Anglo-Saxon or Scandinavian for six hours a day, and that after a year or two the worker gives it up in utter boredom. Such work can be done, as I learn that Ford has discovered, by Slavonic or oriental workers without mental strain for long hours and at low rates of pay. The Anglo-Persian Oil Company have had a similar experience at their works in Abadan in Persia; they find that Persians can work for long hours in their factory in a noisy department where tin-plate oil-containers are made by a mechanical process when an Englishman becomes quickly physically and mentally exhausted. Moreover, can we in any case, however much we rationalise, expect to compete with such enterprises as the admirably equipped cotton factories in China, already referred to, set up by Japanese financiers and worked by Chinese labourers, or the new perfectly equipped boot and shoe and cloth-making factories which are springing up in Czecho-Slovakia?

There may be, however, some temporary increase in our export trade, created by financial or political arrangements, and we may be able to make special terms for trade development, especially in the trade of providing equipment, with some of our dominions and colonies. Nevertheless, such increases, which may possibly be on quite a large scale, cannot from their very nature be permanent. It will, I feel sure, be entirely out of the question so to develop our overseas trade as to absorb permanently more than a small proportion of the workers displaced by rationalisation.

There is no reason to be depressed by this conclusion;

on the contrary, it should be recognised as a hopeful feature. It will be so recognised if we grasp two essential facts: the first, that overseas trade, other than what is really complementary, though it gives employment to shippers (48), traders and financiers, is more likely to do harm than good to our national life as a whole; the second, that our future depends on our withdrawing as soon as is practically possible from such purely competitive overseas trade as is damaging to our home trade. Thus we shall limit overseas trade in the main to complementary trade (the exchange of what is needed in each case by the importing country), a form of trade beneficial to both nations. We shall then be buying what we want, and not, as we do now, what we do not want. Concurrently to meet the problem of employment and wealth production, we must, as is explained later, develop our home industries and agriculture, and so our home trade.

Let us consider and analyse a few specific facts that throw light on this problem of rationalisation, with relation to production and consumption.

"A machine to-day," writes Professor Soddy, "can embody tens of thousands of horse-power, every horse-power equal to the work of ten men, and that machine can work eighteen or twenty-four hours a day, so that it can produce as much wealth to-day as three hundred thousand men in the time of Adam Smith." Does anyone suppose that we can develop consumption in proportion?

Now, let us take something simpler. Miss Margaret Bondfield, when in 1931 she was Minister of Labour in the Labour Government, pointed out in a speech in the House of Commons (49) that certain banks had dismissed 311 of their ledger clerks, to be replaced by 86 women who did the same work with calculating machines, and one hears similar stories from other sources. Indeed, offices are becoming rapidly mechanised by every sort of calculating machine; and machines are coming into use for every possible purpose, even for stamping and opening envelopes. In the Civil Service we were told early in 1931 that hundreds of officials had already been displaced, and according to the *Daily Telegraph*, "When the transition is complete, it is estimated that there will be a saving of human labour to the extent of 500,000 hours a year. The annual saving in wages will be about £13,000 per annum." (50) It did not seem to occur to the writer of this phrase "saving in wages" that there would be no saving: for the persons have to be supported in any case, and probably by the state. Does anyone seriously suppose that banking, the Civil Service or commerce are going to be extended to such an extent as to absorb these men so dismissed?

A further simple example of the effect of rationalisation was given me in the year 1930 by a gentleman at that time actually employed in rationalising industry. (51) He had had to deal with three factories, engaged in a certain branch of the paper trade, for the most part, I understood, at work on government contracts. He had closed one, dismissed about four hundred men, amalgamated the other two, and guaranteed the dividends of the shareholders in the closed-down factory. The same work was being done by the men who remained. There was no reason to suppose that we were going to increase the demand for these goods in proportion.

One might give innumerable examples of how intelligence, new mechanical appliances and amalgamations create unemployment, but it is unnecessary—the facts are clear; we can, however, take a special problem of the press. The rationalisation in the form of amalgamation of newspapers that is recorded from time to time has caused serious unemployment amongst journalists; here and there a man gets back to work, but on the whole it seems probable that the reduction of the numbers employed in journalism is permanent.

As a result of rationalisation, productive power has increased and is increasing enormously: theoretically, we can increase consumption indefinitely; nevertheless, there can be no possible immediate increase of consumption that would bear any relation to this increase of productive power. Moreover, to take our own special national problem, the only increase of consumption that can be hoped for has to be found in our own country, for, as has already been made clear, other countries will shortly be able to supply most of their own needs, or obtain them from cheaper sources. But home consumption is being stultified for a very simple reason. Rationalisation puts workers out of employ; therefore their power of buying is reduced and home trade goes down automatically.

The general effect of rationalisation, in this country at any rate, is therefore to increase production and reduce consumption.

This problem of over-production can be illustrated from a slightly different angle. There has been in recent years production in excess of the demand in cereals, oil, rubber, cotton, coffee, tin and no doubt other commodities. The condition of the wheat trade, as it was in 1930 and 1931, illuminates the general situation, and is perhaps the simplest to explain. The farmers and peasants of far too many nations had concentrated their energies on growing wheat for the export market; they had no doubt been specially tempted to do this by the British open market, the outcome of our free trade policy. But apart from this, and as a result of the world-wide system of unregulated trade, the main exporting countries had produced more than the demand justified, and had at that time surpluses in hand. The United States of America, Canada, the Argentine, Australia, the Baltic nations, the peoples of the Danube Valley were all concerned. Conferences in Rome and London considered the problem in the year 1931, but what could they do? Two suggestions were made; one came from the United States and was in part adopted. Let America, it was said, get rid of its surplus, either by burning it or by giving it or selling it cheaply to China. (52) Another suggestion came from Rome. Let the producing nations make a combined effort to persuade oriental nations to consume more wheat. The talk went on whilst farmers and peasants were being ruined all over the world—by over-production, the outcome of improved methods and unregulated trade. It is difficult to predict what will happen if there is no regulation of production. Wheat production may be reduced temporarily, with, as a result, widespread unemployment or under-employment amongst wheat producers. Prices may then go up, possibly to fancy figures, and this in its turn may be followed in a few years by another period of over-production.

3. BRITISH INDUSTRY

During the XIXth century wealth production went on rapidly. We really did become the workshop of the world. There were periods of great prosperity, and we produced wealth for our own nation, of which the greater part went to the financiers, traders and captains of industry. It was also true that in industry as in agriculture competitive trade tended to force down the price against the producer and up against the consumer, but mechanical and other inventions reduced the cost of production so rapidly that industries in this country at any rate were not degraded in the same way as agriculture. The main trouble was the uncertainty of trade, but industrialists were able in times of difficulty to meet the situation so created by putting the workers on half-time or even closing down their workshops and dismissing their men temporarily; a partial solution impossible for agriculturists. There was in the XIXth century none of the decadence in industry that marked agriculture; it was not until early in the present century the tide began to turn. But even in the last century the uncertainty that arose from unregulated home and foreign competition had diverted our industrial leaders, as has been already pointed out, from their obvious business of providing homes and the other necessities of life for the people of their own country in order to join in the struggle to secure trade, or to invest their money in competing industries abroad, and in this century the tendency to invest abroad increased rapidly.

Concurrently came, as has been already pointed out,

the development of selling agencies; intelligence, energy and organising ability were turned to the arts of sale. The world was overrun by travellers and other agents. Our friend the pig-dealer's tout is only a humble example of a widespread class; whilst recently even the Royal Family has used its great gifts and personal charm to help in the development of trade. At the same time every sort of ingenuity, skill and even genius was devised to create attractive advertisements. Even if people did not want the articles that are produced, they had to be hypnotised into buying them. Even if they could not pay for the goods, the consumers were beguiled into mortgaging the future by a system of payment for goods by long-deferred instalments. Advertising, like foreign trade in its early days, was looked upon as a crusade. Enormous sums have been spent in the advertisement crusade; and these sums have to come out of the pockets of the producers or consumers, or, if the trade thereby increased depends on rationalisation, it may come ultimately out of the pockets of the state in unemployment grants. It is well, when you read an attractive advertisement, to remember that you may have to pay for it.

There was one curious and somewhat unexpected result of this advertising development. The growth of advertisements put the press into such a position that in normal times the public buys its papers at perhaps a third of the cost of their production, in some cases probably even less: the balance of cost has to be paid by the advertisers. The proprietors of newspapers would be more than human if they did not give at least some consideration in their business policy to the cause of development of trade and the advertisements that go with it. Moreover, when trade dwindles and advertisement receipts come down, the press is liable to be faced by a problem that seems almost unsolvable. The simplest immediate solution appears to be amalgamation, which causes on the one hand unemployment amongst journalists, and on the other hand the disappearance of newspapers of independent opinion. Stunts and slogans, good, bad and indifferent, then get their chance and exercise an undue influence. Reason is liable to be squeezed out: it may be too dull for modern journalism. This is in itself a disaster; but it is doubtful if amalgamations will ultimately solve the problem. In the future it may be that newspapers will be reduced in size and raised in price, or else most of our national papers may disappear altogether, for advertisers may favour the local press. Papers with a smaller circulation in a smaller area may take their place. If so, the result may be good. We cannot say what will happen, but the period of change causes suffering amongst journalists, and tends to deprive the public of independent thought.

The general effect of our trade system on wages and workers in industry must be referred to, though it cannot be discussed at length. Imbued with the war spirit, the industrial workers have combined, and for about a century have fought for better wages and conditions. Sometimes they won and sometimes they lost, but even when they won their employers were on occasions able to recoup themselves by rationalising their industries and dismissing a proportion of their employees.

4. BRITISH AGRICULTURE

We have already learned something of how competitive trade reacts on agriculture and turns the business of production of the people's food into a gamble.

We have now another point to consider. Farmers all over the world are peculiarly the victims of unregulated trade, and are especially so in this country. It is important to realise why. The cultivation of the land and care of stock is a continuous occupation: its problems have to be dealt with day by day; they may even continue through the night. A farrowing sow or a sick beast may keep a man up all night, and if he is not careful to keep wide-awake, he may lose either. A market-gardener may have to be up at incredibly early hours to get his produce to market; if he is late he may lose the market and waste the result of weeks, even months, of work. In the life of a farmer or market-gardener in this country there is little rest. Everything has to be watched continuously. He has always problems before him and has little chance of freedom from work, or at least freedom from thinking of his work. He has also always to be making important decisions. A farmer gets up in the morning; the day is doubtful: should he cut his hay or not? He decides to do so, and in the afternoon a heavy rain begins which lasts for days: his hay lies rotting. Or he decides not to do so and it is fine: he loses at least a day. Moreover, he does not close down his business as a factory is closed when the day's work is done. Whilst, if prices fall and trade is bad, his problem is entirely different from that of the factory. The manufacturer of boots may find from his travellers at the end of the week that orders are not coming in: in seven days at any rate he can put his men on half-time, he can even close down altogether. If a farmer were to take a similar course, his stock would suffer, and in a very little time his land would be overgrown with weeds, which might well take years to eradicate. He has also both to look back and to plan out for years ahead, considering what crops were grown in the past and in what respect they have exhausted the land, and what should accordingly be sown or planted in the future so as to secure the best use of the soil.

Under modern conditions farming is the most arduous but yet the most fascinating business in the world. Its constructive character creates a zest that is almost unique in modern life; and the workman on the land, with rare exceptions, joins in this interest and acquires this zest. The pride that a farmer and his workmen feel in producing a beautiful field of corn ripe for cutting is a unique thing—one of the priceless things in life; but this happiness turns to bitterness when the farmer finds that he cannot sell his crop, save at a heavy loss.

It is then important to realise that most farmers are, from the very nature of their occupation, immersed in their constructive work of production round which their thoughts centre. To be a good farmer a man must either have the intuitive knowledge of the true peasant or else creative genius. The modern English farmer is a man of rare constructive ability. He may not look like it, but he is in his way both an artist and a scientist.

One has then to realise that men of the farmer's character

and so immersed have, with of course some exceptions, no time, even if they have ability, for regulating the sales side of their business. It is this personal characteristic that makes the farmer the peculiar victim of the trade system, with the result that all the world over (model Denmark included) agriculture tends to become what is relatively to the wealth of the world a sweated industry. This is perhaps the most remarkable result of the Trade War: that the basic industry, the production of food, should, notwithstanding relatively high prices for the finished product, have to be carried on on conditions that never give security, and often fail to give even a bare living to the productive worker.

Nevertheless, the British farmer is always being criticised by politicians, economists and other clever persons, who understand neither his character nor his problems, and attack him for his foolishness. It is not the first time in the history of civilisation that special ability has been confused with folly.

There is another element in the farmer's life. When he leaves his actual work of cultivation his mind is occupied and worried by the marketing system that the competitive trade system has created. He knows he has no control over either distribution or over prices, and that whilst, since he has to plan out for years ahead, it is essential for his success to have a secure market and a standard price, his markets are in fact uncertain, and prices vary every day and in some markets even every moment. What worries him is that he has no control over this problem and never can have. Many men actually break down under this strain. A few find their own solution of the difficulty. They become attracted to dealing, and so benefit by the worries of their fellows.

The effect of our trade system on the special position of agricultural labour in this country may also be referred to. When, nearly a century ago, free trade, external and internal, were crystallised into a national system, what has been called the "dominance of the dealer" became firmly established, and prices to the producer were forced down below the economic level; the farmers at that time re-couped themselves by keeping down wages to a starvation rate and the labourers suffered. In the latter years of the XIXth century, when the free trade system had forced prices down to an even lower figure, labourers and farmers suffered together. In the recent years, when labourers' wages have risen and are protected by Acts of Parliament, the suffering falls on the farmers, and, so far as they are able to do so, they mechanise their farms and dismiss a proportion of their workers. The labourers always suffer.

We can further examine some other aspects of the effect on agriculture of this system of unregulated trade. The complete uncertainty in our markets and variation of prices results in either under-development or over-development, and both may be acute evils. We get both on a large and on a small scale. Recently, as has been already noted above, there has been special over-development in the business of corn-growing in many countries, and there is also permanently over-production of bacon, eggs and butter in Denmark. On the other hand, there is under-development of almost all forms of food on a national scale in Britain; we also constantly get both over- and under-development on a smaller scale in the case of fruit and vegetable crops in this country.

Both over- and under-production are at times disastrous to the agricultural workers of all or almost all countries; and though traders, financiers and transporters concerned with agricultural produce may on occasions benefit, in recent years they also have suffered. That is the hopeful feature—for they also are beginning to think.

This, then, is the main economic outcome of our trade war on agriculture to-day: over- or under-development. The position is so difficult in this country that the greater part of our agriculture might well have been destroyed long ago, were it not that our people are extremely persistent, whilst nature is extremely prolific. What this productivity of nature means is often overlooked. A grain of corn may produce a hundredfold and even far more; an apple-tree may give many bushels of apples; a sow may have two litters of pigs in a year, and each may be a dozen; a cow may give seven hundred gallons of good milk in a year, and, in special circumstances, a far greater return. Each of my little lot of pullets provides me, on on average, with about a hundred and sixty eggs in a year, and skilful hen farmers get much better results: a few shillings' worth of seed provides me with vegetables all the year round. Were it not for this productivity of nature, combined with the advantages of a moist temperate climate and a good soil, all English farmers would be ruined by the trade war. As it is, they are half-paralysed. Thus has come under-production on a large scale and unemployment, with dismissals and reductions in the number of workers on a far larger scale than any other industry, with loss to the country, which in the year 1931 was computed at £250,000,000 a year, or £5,000,000 a week, and was possibly far more. (53)

5. INSECURITY OF LIFE

There is another aspect of the system of unregulated trade to be considered. It destroys security of life and of employment amongst our workers. By immense skill and genius a financial group captures, for example, the soap trade, and a number of firms are bought or squeezed out. Unemployment is certainly created, temporarily, probably permanently. Or multiple shops capture the retail business of selling medicines and other relative trades: this time chemists are squeezed out; concurrently every effort is made to persuade people to take drugs and patent medicines, on which there is admittedly an enormous margin of profit. Or we build laundries all over the country and put out of employ or transfer women workers. We move our factories and workshops, not only, as has been already pointed out, to other countries, but from one part of Britain to another, and the people have to drift about. Sometimes we make an immense mistake. We have drawn our people out of agriculture and other industries into the coal and iron and steel and other related trades, and crowded them into towns and industrial villages, only to find that the market for these products has disappeared, and we have hundreds of thousands of people permanently unemployed. Many people undoubtedly become rich, some inordinately so, in the course of this industrial rearrangement, and as a result of this and other causes the luxury trades are rapidly over-developed. Thus they have taken a position entirely out of proportion to the industries that deal with the necessities and amenities of life. Incidentally, it may be noted that the luxury trades lend themselves specially to advertisement.

It is useless to reply to the description and analysis given in this section that it will all be right "in the long run." I can only reiterate that we do not live "in the long run": we live in the present.

I must also reiterate that I am not here concerned in proving any particular theory, but in trying to give a vivid and I hope fairly accurate picture of certain effects of uncontrolled trade—the Trade War. What we have to do is to face these facts, to recognise that a large part of our national troubles is the outcome of our struggle for trade at any cost. Then even if we feel that competition will always be a feature of our civilisation, we may yet recognise that it cannot be made the basis of a prosperous industrial and agricultural life. Then we may well go on to consider whether this competitive struggle, with all the energy, skill and ability by which it is supported, cannot be transferred from the pursuit of commerce for its own sake to the work of securing the necessities and the amenities of life for our people at large. I suggest that such diversion of energy is not going to be so difficult as it at first appears, provided that we face the facts.

H

PART III

VARIOUS MATTERS FOR CONSIDERATION

VARIOUS MATTERS FOR CONSIDERATION

I. THE PROBLEM OF PRICES

Nothing in our social life is more ridiculous than the so-called "vacillating price system" into which we drifted in the XIXth century, and have been vainly trying to operate ever since. It is so entirely unsystematic that it is not even entitled to be called a system. Something has been said of it incidentally in the earlier part of this essay; it cannot, of course, be dealt with here at length— it is too big a subject, but it may be referred to shortly, for it has direct bearing on our industrial and agricultural policy.

It is obvious to anyone who has given any real study to the relation of trade, industry and agriculture to civilisation that whatever price system we do adopt, our aim should be to maintain prices at a fair level as between producer and consumer. This has been recognised for centuries. Cobden, it will be remembered, grasped this point, and argued that if free trade were adopted prices in corn, at any rate, would be automatically stabilised at such a level. He was wrong in his argument, and no doubt free trade economists knew that he was wrong, for whilst Cobden was recognising the evils of vacillating prices in corn, other free traders were justifying fluctuations of prices. Prices varied, said the free trade economists, with the law of supply and demand;

no attempt should therefore be made to control prices, since under the beneficent action of this law supply and demand would be automatically equalised. This theory and the whole theory of the law of supply and demand on which it was based are now known to be untrue. The economists were nevertheless, in those days, very emphatic about this so-called law; whilst free trade was only a moral law, comparable, as Bright had suggested, to the Ten Commandments, and therefore easily broken, the law of supply and demand, they asserted, was as the law of gravitation which never could change. It controlled trade and industry whatever social and price system we adopted. There was an inconsistency between the two views advanced by the two sections of the free traders, but in their political propaganda this did not matter, for they thus secured the support of two groups: the practical men who wanted fair, steady prices, and the intellectuals who thought that prices should and must vacillate; both groups supported free trade, though for opposite reasons. Incidents of this sort, when two conflicting theories are put forward by the same party to secure the support of persons of two schools of thought, (54) are a commonplace in political life.

It is important to get some idea of what is meant by the law of supply and demand, for there is a grain of truth in it; moreover, it does help to suggest a solution of the price problem, a matter of real importance, for on its solution the whole of our economic life depends. The case put forward, then, by the economists was substantially as follows. As supply and demand varied, prices went up and down automatically and in due relation:

if there was a small surplus, prices would go down slightly; if there was a shortage, prices would go up, also in proportion, and as a result supply and demand would automatically adjust themselves. Of course it is almost, but not quite, a myth. A small surplus, a reputed surplus or even fear of surplus of some article, especially if it be perishable, may throw the control of the market into the hands of buyers, and if the surplus be accompanied by a sufficient scare amongst producers, may bring prices down with a run—perhaps to nothing. This actually happened in this country in 1923 in the potato trade, when there was only a small surplus, and yet prices fell to next to nothing: whilst in the wheat trade in the years 1930 and 1931 a world surplus, computed at 14 per cent., (55) drove prices down about 50 per cent. In the same way a shortage, or fear of shortage, may send the price up to a fancy figure. Let us illustrate variations of prices by a very small example of what has actually happened in Covent Garden: it is an extreme case, but it is very illuminating. When on one occasion the Covent Garden market opened in the early morning, English hot-house grapes of a special quality were sold at 4s. 6d. per pound. In an hour they were down to 1s. 6d., and before the market closed they were up to 10s. 6d. (56) What was the exact cause of these variations we do not know; but the fall was probably due to a belief that there was a glut in the market, whilst the ultimate rise came, no doubt, from fear of shortage. Supply and demand had something to do with it, but obviously there was no absolute definite relation between supply and demand and prices. Prices are in fact influenced by other things, of which the most important of the immediate causes is bargaining power, with in the background and on a large scale the varying values of money. The story of the gamble in tomatoes described in an earlier section is a good example of how bargaining power may actually defeat the tendency of surplus supply to lower prices, for in that case, notwithstanding the surplus, retail prices of tomatoes in London went up to a high figure.

All we can say about supply and demand is that it is one of the causes of variation of prices, and if we are going to secure steady prices we have to see that so far as possible supply and demand are equalised. The countryman realises this when he tries in vain to ensure, against market day, that nothing more goes into the market of his country town than the market will take up. Equalisation of supply and demand is sound in principle, and the principle has universal application. To maintain a price level we have then to aim at a continuous equalisation of supply and demand in the home market, and as far as possible in the markets of the world.

We have already seen that bargaining power directly influences prices in small matters. But this influence is universal. A buyer with money in his pocket or with the banks behind him is in a position to buy cheaply—especially if he is not in a hurry—from a manufacturer with a heavy overdraft and stocks in hand that he wants to clear for cash. Whilst if the manufacturer is doing well, has ample funds in hand and feels secure in his market, the boot is on the other leg; the balance of bargaining power is in his hand, and he can hold on and may get a better price than he otherwise would. The differences in price may seem small, but they may be just the difference between a loss and a profit. A tendency of banks to finance merchants rather than producers has, it may be noted, a special influence on prices, as it puts bargaining power into the hands of the merchants.

We have also to consider the relation of high finance to prices. It arises in this way. Under the finance system which we have adopted, money has not a permanent unchanging value—is not a fixed measure. It is not like the foot-rule of the carpenter, a fixed length, but is a thing that itself varies. A user of our variable money may be compared to a carpenter who found that the length of his rule varied from day to day with, one may suggest, the weather. The scale of the money measure varies, indeed, as inconsequently as the weather. A pound to-day may be 15s. to-morrow and something quite different the day after. Of these variations more is said later: the immediate point to realise is that variation in prices is not only due to variations in supply and demand and to bargaining power, but to change in value of money. This change in value of money creates complete uncertainty and vitiates all business transactions. To take one example. Falling prices arising from a rise in value of money frighten the producer, who hesitates and is inclined to reduce production and so employment: a rise may have the opposite effect and increase production and employment. Both effects may be temporary, but they are none the better for that. These temporary changes may be disastrous in their results on the lives of men and women.

There are two other disconnected points relating to

prices that may well be referred to here. The first is this. British free traders became obsessed by the idea of cheapness. "To buy in the cheapest market and sell in the dearest," was, so Cobden told the House of Commons on one occasion in 1846, "one of the ways of carrying out to the fullest extent the Christian doctrine of 'doing to all men as ye would that they should do unto you.'" Is it? There may be personal advantages to private individuals in buying at as low prices as possible, but from the national point of view serious evils may arise. Take, for example, the case of the Canadian farmers who were forced in the years 1930 and 1931 to sell wheat to us at a very low figure, in some cases far below the cost of production. The first result was that many were brought to the verge of poverty and some actually ruined; then, as an ultimate effect, the impoverished Canadian farmers are either unable to buy our goods in return, or can only take goods at very low prices. Consideration will show that there is a very real danger in making cheapness a fetish. High and low prices are both evils for both the producer and consumer—steady middle prices appear to give best results. The second point that we ought not to lose sight of is this. The effect of competition on prices is not necessarily to lower prices; it is varied, but in all cases it appears to create vacillation. In our overseas trade, with a system of free imports, the general effect of this competition is to lower wholesale prices. On the other hand, the general effect of internal free trade which puts power in the hands of intermediaries is to lower prices as against the producer and raise them as against the consumer.

It is not possible to explain the intricacies of price variation or of the relative value of different scales of prices, and their relation to the value of money in a popular essay like this. But the two main points that we have to get clear in our minds are these: the orthodox economists were entirely wrong when they said that prices were controlled by a beneficent law, a law of supply and demand, and also when they argued that our primary object should be to buy in the cheapest market. We may conclude from this that our national policy of leaving prices to take care of themselves is based on a fallacy; for the general result is that prices vary, not only with variations of supply and demand and with the bargaining power of individuals, but on a large scale at every breath of national thought and action. These variations give the expert dealer or financier many opportunities to profit by such changes, and so become rich at the expense of the productive workers and consumers. (57)

I venture to suggest that in dealing with prices, our national policy should have two objects; the first to stabilise the value of money, making it a real measure of value; the second, to introduce as far as is practical either definite standard prices for at least the main articles of our food supply, to be fixed at a fair figure as between producer and consumer, or if that may not be immediately possible, to secure that prices are stabilised within definite limits. What the standard price should be and how it should be fixed are problems that require much more investigation than they have yet received. Many people in this country are confused by the idea, cleverly popularised in the past by our political and economic leaders, that what is called the "world's price," a strangely meaningless phrase, is the right price. (58) The late Woodrow Wilson, President of the United States, had a different idea; some years ago he gave a definition of what he called the Just Price, which is certainly illuminating. "By a just price," he said, "I mean a price which will sustain the industries concerned in a high state of efficiency, provide a living for those who conduct them, enable them to pay good wages and make possible the expansion of their enterprises which will from time to time become necessary." But we want to go a little farther than this; we have also to consider, for example, the effect of particular prices on problems such as production of wealth and unemployment, and endeavour to secure as between producer and consumer a scale of prices that is also beneficial to the nation at large. (59) I am inclined to think that in food products, at any rate, this will not be found so difficult as it appears to those who have not made any special study of the question.

2. UNEMPLOYMENT AND POPULATION

Notwithstanding all the talk about unemployment to-day, there is very little clear understanding of the facts and the problem.

Let us consider first the facts so far as they affect our own country. Week by week figures are published in the press of the number of the unemployed who come under the Insurance scheme. The average figure in the year 1931 was about 2,650,000. But there were very large numbers of unemployed not included in these figures. Many middle-class workers, small men with businesses of their own, all workers on the land including gardeners, regular and jobbing, and domestic workers; then there were numerous very poor people who either did not get any form of relief or were helped by the poor law or private charity; there were in that year, on an average, about a million individuals receiving relief from the poor law, of whom perhaps a tenth would have been normally in employment: (60) 3,000,000 is a low estimate of the average number of individuals unemployed in any particular week in the year 1931. There is also to be considered the fact that there is a constant shuffle of employment: a very large number of individuals are always going in and out of work, and those who are insured only come intermittently "on the dole." Figures published by the Ministry of Labour in 1928 go to show that over a period of two and a half years from October 1923 to April 1926—when the average number of workers "on the dole" was round about 1,300,000—the number of individuals who actually drew unemployment benefit at one time or another was about 4,000,000. (61) This suggests that if we are to get a true picture of the number not in constant employment in 1931 we might have to treble the figure of 3,000,000. Many of these individuals would have been unemployed for only a few weeks in the year, and we may leave them out of the account. But even then, if we make a conservative estimate, we may say with a reasonable degree of certainty that 6,000,000 of our workers, if they found work at all in the year 1931, were only intermittently employed. This, of course, is only the beginning of the trouble; there are dependents of one sort and another involved: wives and "unmarried wives," to use a phrase that comes from wartime; children, legitimate and illegitimate, old parents and others; if we allow only one dependent to each of the unemployed this will give a total as high as 12,000,000. Twelve million people definitely affected by want of work is a serious matter; the actual number may be far greater. It is not, indeed, unreasonable to suggest that half of our population were in 1931 living under the shadow of unemployment. What unemployment means in the actual daily life of the people, with all the poverty and anxiety that it involves, is hardly realised by our intellectual leaders. It is true that many unemployed individuals accept the situation with equanimity, but such equanimity is, I fancy, the exception rather than the rule. Moreover, from the national point of view it is an appalling tragedy, even if it is only considered from the point of view of the loss of national wealth. The position would have been even worse were it not that between two and three hundred thousand people were said to be directly and indirectly employed in that year on what are substantially relief works (62) —a state of things that can hardly be continued indefinitely.

We can now leave our special British problem to a consideration of other aspects of the question, and in doing so we may well clear out of the way two special points that confuse the minds of many persons—they are the effect of the war and of population on unemployment.

It is certainly a mistake to suppose that our present

unemployment arises automatically out of the War. It is true that the War destroyed what may be described as "the goodwill" of our overseas trade. It stopped the stream of trade and broke down the steady flow of those "repeat orders" that are so easy both to give and to execute. It also, no doubt, caused the peoples of other nations to see that in a large number of cases they could provide for themselves or secure from other sources many of the things that they were accustomed to buy from Britain. But all this would have happened in any case, and indeed, had there been no war, the series of incidents that have created the present unemployment might well have occurred some years ago, with results similar to those from which we are suffering to-day.

It is further constantly suggested that unemployment arises from the size of our population. Rapid increase of population certainly creates special difficulties, and if population becomes stable in this country, as it is likely to do, it will help to stabilise the position which is always a great help to any constructive effort. Moreover, if it were possible to reduce our British population to, say, 30,000,000 by some scheme of mass colonisation, it would again simplify the problem, for we should then have no difficulty in becoming a substantially self-supporting nation. Nevertheless, to-day there is undoubtedly plenty of work in this country for our workers, and the size of our population is not the basic trouble.

The removal of these misunderstandings leaves the way clear for a consideration of the question at large. To do so intelligently we must put aside the idea that there is some natural cure for unemployment, that there is a "Law of Nature," as Adam Smith appeared to have believed, that creates employment, that brings work and worker together and starts them doing it.

This done, we have to consider the problem anew, and for this purpose it is well to go back to origins—indeed, when one goes back to the origins of evil it seems that by some strange philosophic law the problem becomes clear and the solution emerges.* Perhaps this because when one views the problem as a problem of all time, one accidentally secures objective vision,† which is the essential of understanding. In this present case, what first emerges from the study of origins is that the problem is not really a problem of unemployment, but a problem of providing employment. That the cure of unemployment is employment in constructive work is indeed constantly and strangely overlooked. When that outlook is attained the matter is greatly simplified.

To turn then to origins: even in the most primitive times Nature took no part in solving this problem; she was neutral or even hostile. It was not her job: the savage had to find his work and do it, and that is really what the community has to do to-day, for the essence of the problem to-day is that the individual can no longer, in a large number of cases, find his own work. This was recognised in medieval times—in fact if not in theory—for in those times, in this country at any rate, the community helped to bring work and worker together. If in village

* This is certainly true of the trade and housing problems, and will be, I think, of the financial problems when they are examined.

† That is, one sees the problem in time as well as space.

life a man wanted to build a house, the village community found him the site and the wood or other material free of cost; if a craftsman in the town wanted to set up a workshop, his guild, in suitable cases, provided him with credit without interest, so that he might start on his own account. When I was engaged in reconstructing civilisation in Central Europe, we had to revert to both these primitive methods; thus we created not money but wealth —in fact, houses and food.

The introduction in the course of the growth of civilisation of private property in land and private control of finance has no doubt made the general position more difficult, since both, whatever their advantages in other directions, are liable to create a barrier between the work and the workers. Would-be workers can no longer go to the community for land, for materials or for credit. Rationalisation obviously aggravates the difficulties of bringing workers and work together, and so does freedom of trade and the competition that goes with it. Either may at any time go far to reduce employment in some branch of industry or agriculture, or actually destroy it and throw numbers of people suddenly out of work. Our finance system, which provides credit only for work that is financially profitable, does not help the situation as it might. To-day we see all this clearly.

To solve the unemployment problem we have, then, to bring together the workers and the work that needs to be done, and also to set the workers to work. This is the problem that faces us in this country to-day. We have to bring the horse to the water, but he has also to be made to drink. The parallel is not very exact, for all

I

workers have not the natural desire for work that horses have for water. To solve the problem we also have first to find for the worker work for which he is adapted, or else to train him; this done, we have either to compel him or else to make it well worth his while to do it: in this country we are in a peculiarly fortunate position for dealing with the problem, for there appears to be much work waiting to be done to absorb our unemployed, and there is no fundamental difficulty such as is to be found in other countries in starting them upon it. It appears, to illustrate this general position, that there were in 1931 nearly 300,000 men out of employ in the building trade, and yet there was need of somewhere about a million houses for the workers; agricultural labourers were standing about in the villages and displaced country-men in the towns, whilst much land was either actually uncultivated or under-cultivated. School-teachers were unemployed, and owing to the size of the classes, children were not securing the teaching they needed. Much of this and of other work requiring to be done was wealth-producing work, in a large part of the cost of which we were actually involved, since the men and women who might be doing it were kept idle, and drawing pay in the form of doles from the pockets of our other citizens.

It is very important to realise the point already emphasised, that there is no law of nature that naturally creates employment. It is perhaps not too much to say that the present excessive unemployment arises automatically out of the form that our civilisation has taken, for civilisation has first put up barriers between the work and the workers, and thereafter has fostered unemployment by the introduction of competitive trade, rationalisation and a special and peculiar finance system. Moreover, we have clearly to realise that to solve the problem of unemployment we have to make a conscious national effort to bring work and workers together, and to ration out the work amongst them, for it is not rationalisation that is required but rationing.

3. THE PUZZLE OF FINANCE

From the lobby of the House of Commons, where our legislators relieve the strain of life by exchanging amusing and sometimes scandalous anecdotes, comes a story, *Se non è vero è ben trovato.*

One summer not so very long ago a group of Members of Parliament was formed to study the puzzle of Finance. The great financial authorities addressed it in turn; each, so it is said, told a separate story, and none was able to answer a simple question in a simple manner.

To those of my readers who have read the Foreword of this book, the reason is obvious. The distinguished authorities had all looked through the peep-hole of the Social Kaleidoscope and had each seen a different picture, from which they had drawn different conclusions.

The fact is that the problem of finance requires a new analysis, for it really remains unexplored. Just as the problem of trade was permanently obscured by the teaching of Adam Smith and his followers, so the problem of finance is obscured, it appears to me, by old theories, probably in this case those first put forward by Calvin and his followers, and handed down from generation to generation without real reconsideration.

We want to deal with finance from an entirely different outlook, in the same spirit as I have endeavoured to study trade. I think then we should find that, though the problem is complicated, the solution is simple. All this cannot be gone into here. For the moment I suggest that we may rightly look upon our present financial system as a cankerous growth that has spread through the social system. This growth has resulted in a large number of specific evils, of which the fundamental is that finance has become the controller of life instead of its servant.

It may be useful if I turn from generalisations and endeavour to examine certain specific details of the present system, leaving complete analysis to a later work. This course will, I hope, at least help readers to realise the importance of the subject and greatness of the evil.

We will take first the so-called "gold standard." There cannot and never could be a gold standard in any real sense of the words, for gold has no real fixed value in itself, whilst such doubtful value as it has, since it is in some sense a commodity, varies with supply and demand, and also with what we have called bargaining power. We cannot fix a standard of value by relation to something that has no fixed value. The belief that gold has real value is, I suggest, an obsession—it has no support in fact. Nevertheless, the "gold standard" as a practical proposition does seem in the past to some extent to have worked effectively; it was believed in, and this belief—an "act of faith"—controlled the situation, for in the life of civilisation faith is all-powerful. Thus was created in the imagination of the world a fancied value, and this may continue to cause gold to be accepted, even by future generations, as if it were a thing of basic value; as has been said, "financiers think in terms of gold," and so have and still do many other people. Nevertheless, a time must come when the faith will die, and the whole mechanism of finance, if still based on the gold standard, will break down.

It is, then, the absence of a fixed standard of value that lies at the basis of the variations of the value of money that cause so many troubles. It makes it difficult for financiers working even with the support of governments to control money values and so price-levels; indeed, it is obvious that the financiers and governments often lose control altogether, and are as helpless as the public to prevent changes. This I can illustrate from my own experience. When I was reconstructing agriculture in the White Russian provinces of East Poland in 1922, the Polish mark, which had started a few years before at the nominal value of approximately a shilling, had gone down to a thousandth part of a shilling, and afterwards it went down until I was told one could buy about a million marks for one shilling. To meet these variations I adopted a system of my own. I paid wages in food—fat bacon and flour—and I gave the customary presents to the officials whose support I needed (what in this country would be unkindly called bribes) in sacks of corn or potatoes, or ploughs, spades or other tools, all of which were urgently needed and so of price value. But such simple methods, in fact a return to barter, cannot be applied on a national scale. Germany suffered in the same way as Poland, and similar, though not quite so catastrophic, changes happened to the value of the franc in France.

Changes of money values happen constantly on a smaller scale. In the years when England readjusted her financial system after the year 1920 and ultimately in 1925 adopted what is called the "gold standard,"* prices came down with a run: some people were no doubt ruined and many lost largely; but others, especially financiers, seem to have gained. Following the events of 1931, when Britain went, as it was said, "off the gold standard," prices may be expected to go up: certainly the prices of foreign goods may rise. Such changes create uncertainty and many other difficulties—a fall of prices specially frightens and paralyses manufacturers and other productive employers who hesitate, and are inclined to reduce production and so create unemployment. A rise has the opposite effect. If all wages and prices, rents and interest charges varied automatically as value of money changed the evil might be less, but they do not.

The absence of a fixed value also makes possible the dealing, we might call it gambling, in gold and money values that goes on all over the world. To take a small example. At the time of the variations in the value of the franc in France, thousands of people in this country were speculating day by day: putting their money on French francs as other people put it on horses. Nations, financiers and bankers, and also speculators pure and

* What the "gold standard" really implies is that countries who recognise it will exchange their currency at a standard level.

simple, take part in these dealing and gambling transactions, which seem to have been going on for many
years. It appears that so long ago as 1896 a case occurred
in which a group of American financiers started opera-
tions by selling British securities heavily, and buying
United States bonds and shares. The next step it appears
was to draw from the Bank of England sums equal in all
to £11,000,000 in gold and ship it to New York. The
transfer of this gold caused a fall in the prices of 350 of
our representative securities, equivalent to £116,000,000,
whilst the absorption of this gold caused a corresponding
rise of American securities. No doubt the final step of
the financial group was gradually to buy back British
securities at the lower value, and sell the American ones
at the higher. The ultimate result of the transaction is
of course not known. (63)

Variations in the value of money have somehow to be
brought to an end. We have to fix the value of money.
It is suggested that this can be done effectively by insti-
tuting what is called a "managed currency" as appears
to have been done successfully in Czecho-Slovakia after
the War: but to me there appear to be two more direct
ways of doing this, either by fixing the value of money
in relation to something of real value, if that be possible,
or by fixing the prices of the main essentials of life. The
result, I think, is the same in either case. We might try
the latter in our own country at any rate: that is, we
might fix money through its relation to the value in this
country of the main commodities we deal with. Further,
for the purposes of world trade it is desirable to fix the
value of money absolutely as between different countries.

This may be impossible: but it is an ideal to work for. A more practical and immediately possible proposal is to fix the value of money within the British Empire. Indeed, if we are going to develop complementary trade within the Empire, this is a first essential of success.

A further practical question that may be considered with advantage is the problem of "credit."* Credit, using the word in its popular sense, is needed for wealth production, to keep things going whilst the work of production is proceeding. It is needed to bring work and workers together, a point that has been dealt with in discussing the problem of unemployment. The issue of credit for this purpose involves, in the terms of practical life, a proposal addressed to workmen and owners of goods to this effect. If you respectively do the actual work, and provide the materials for the creation of wealth that is needed, you naturally wish to be recompensed out of the results of your labour; that is, out of the wealth you produce. But in the meantime you have to carry on, and during that time you wish to be provided with money— that is, in actual coins, quite a minor feature in such transactions to-day, or else in bank notes, cheques, bills or other securities which stand for money. These notes or other securities can finally be cancelled out against the wealth you have created. This is, in its essentials, what lies behind most constructive business transactions. The point may perhaps be seen more clearly if a practical example is given. Suppose we wish to build cottages. We have in this country clay to make the bricks and tiles,

* The word "credit" is used with many meanings, a fact which creates endless confusion.

and much, though not all other materials needed; we have also the labour immediately available. Credit is needed in bringing the work and worker together, to pay for the materials and to provide wages for the workers, whilst the job is going forward. Without credit, the work and the workers cannot come together. What we have to secure to-day is that credit should be provided to bring our workers into employ and to create national wealth. We have, then, so to adjust our financial system that credit is available for the specific purposes that are needed for reconstruction in Britain to-day, and will be needed in the future. At present credit is only as a rule available for work that is immediately profitable.

There is one other aspect of the finance problem on which perhaps something may be said. What is the fundamental cause of the collapse, total or partial, of so many banks in the United States of America? I suggest it arises out of the fact that the trade war was carried on more vigorously in the United States than in any other country. On this I propose to make some comments, pursuing my parallel of War of Arms and Trade War, that may at first sight appear obscure. The financing of wars has always been an important branch of finance, and we find Venetian finance backing the Crusades and exercising a controlling influence at one time over military policy. Indeed, when there seemed to be no profits to be made by attacking the Saracens, it was due to the Christian bankers' influence that Christian armies sacked Christian cities and communities in order that financiers might secure repayment of their own loans to the armies. (64) From that time up till the great war financiers seem always or almost always in case of war to have protected their own interests, whoever won, and in any case one side won, and they get some of their money back. But in the Trade War financiers are, under modern procedure, engaged in financing all parties, whilst the ultimate effect of the Trade War, when it is carried on to its ultimate conclusion, is that no one wins, for finally all parties collapse together. It is not therefore surprising that the close of the hundred years Trade War was accompanied by a serious financial dislocation such as we have seen recently in this country, and in a more serious form in the United States of America, Germany and Austria, where many banks have actually closed their doors. It is fortunate for this country that the English banks are cautious, and that we have been saved from this catastrophe.

It is hoped that these comments on the puzzle of finance will not leave my readers even more puzzled than they were when they commenced to read them.

In any case the two main financial problems to-day are (1) to make money a thing of fixed value, a real measure of value, primarily in our own country, secondarily, within the Empire, and ultimately, if it be possible, throughout the world. And (2) to secure credit to bring our workers into employ, and so secure wealth for the nation (65).

4. THE DILEMMA OF TAXATION

The main business of the state, according to the theory put forward by our Victorian ancestors of the orthodox school of political thought, was to administer justice and to preserve order at home and abroad. The state, it was argued, should keep its hands off the affairs of individuals: "the law of nature" would then operate freely, the liberty of the individual would be maintained, and all would work out to the general advantage. It is doubtful if this theory is true, but it was accepted, and so long as it was accepted and this line of policy was adhered to, the contributions of individuals to national funds were not very large and taxation was a relatively simple business.

In modern times, under the pressure of democracy, we have adopted a different theory and a different policy. The community, it is argued by the supporters of this new theory, is only a large family, of whom some are rich and some poor. This is not fair. The state, the father of the family, must see that there is a better distribution of wealth. We therefore tax the richer members of the community and make allowances to the unemployed, the aged, the widows and other poor persons, and ease the position of the poorer half of the nation by providing free education and health services. It appeared that in 1931 about one-third of the population benefited directly or indirectly from these allowances, (66) whilst the whole population that lay below what may be called the middle classes benefited by the free education and health services.

Whatever the arguments for or against this particular policy, we have become involved in it; it appears impossible to disentangle ourselves, even if we wished to do so: indeed, notwithstanding the action in 1931 of the National Government, (67) no political party would be prepared to suggest a very large reduction of the incomes and social advantages of a third of the voters.

It might have been possible to work this modern policy of providing funds by taxation from the rich for the poor, had it not been that by adopting a free trade system, which brought food and other goods from other countries, we incidentally threw masses of workers out of employment, and so blocked the creation of the national wealth from which the taxation might have been provided to pay allowances.

The Labour Party, which was in power in the years 1929–31 when unemployment was rapidly increasing, largely as the result of the economic policy of its various predecessors, failed to grasp the point involved in this statement. They assumed, what is perhaps theoretically true, though not politically possible (for the controllers of wealth were able to make a political revolt), that they could increase taxation indefinitely without increasing proportionately the wealth of the nation from which taxation was drawn. They might have succeeded in their policy had they utilised the money collected by taxation in bringing the work that wanted doing and the workers together—that is to say, in employing the workers in cultivating the land and producing the other things that were needed by the people, and so creating wealth. A nation so enriched could have borne additional taxation without feeling it acutely, for it would have been spread over a greater accumulation of wealth. As a result of the policy actually pursued, the country found itself on the horns of a dilemma—if taxation were not reduced, the state might be in difficulty, whilst if taxation was increased, as actually occurred when the National Government came into power, the incomes of a large proportion of the population were reduced, their spending power was similarly reduced, and so more unemployment was liable to be created, and thereafter more expenditure in taxation. It is true that adjustments in the value of money and certain special incidents did, in fact, create more employment, and this increase may go on for a year or two and even reduce unemployment substantially, but this does not affect the general position. The increase of taxation will in itself tend to increase unemployment. The only way out of this dilemma is to get the unemployed back on to work that will produce the wealth from which taxation is derived. (68) From the point of view of taxation, then, as indeed from all other points of view, it is to-day absolutely necessary, if we are to survive, rapidly to increase national wealth, that is, in our case, home production of food and of manufactured goods. There will thus be not only reduced expenditure on the unemployed, but more wealth to divide between the members of the community: we shall thus lessen the amount to be collected, and also lighten the burden of the taxation that remains, by spreading it over the increased wealth made available. Whatever steps are necessary to secure this result should be taken at once.

5. REACTIONS AGAINST COMPETITIVE TRADE

In all countries, including our own, whenever there are leaders who have the intelligence to see that competitive trade may be a destroyer of civilisation, steps have been taken to counteract its worst evils by regulation. The steps so taken are either national or the result of private endeavour, but all have the same object: to regulate trade and so give security of price and continuity of work to the producers, and, as a result, to secure employment and create or maintain wealth. None of these methods is as yet comprehensive enough to give completely satisfactory results. Nevertheless, successful or unsuccessful, they serve as a guide to what can be done in the future.

In most nations of the world some protection has been for many years given to home industries by the method of import duties. These duties may be relatively small in amount and primarily imposed for revenue purposes, but in other countries the duties are directed to prevent importation altogether; for example, in some European countries duties on agricultural produce run up to 100 and 200 per cent. (69) The method of tariffs is perhaps the least satisfactory method of dealing with the problem of regulation of trade; on the one hand, if the import duties are high, and there is no internal regulation of prices, home producers may force up prices within the state to artificial figures; on the other hand, if they are low, they may not afford sufficient protection to give security of price and market to the home producer. But to many nations tariffs have seemed in the past the best and indeed the only system.

The more effective method of regulating foreign trade, which is now coming into use, is the definite control of imports. Control of imports is a more fundamental method than tariffs: it prevents undesirable importations. This control may be exercised either by means of marketing boards, for purchase of both home and foreign supplies, or by limiting the right of merchants to purchase from abroad, so as to secure the home producer a first claim on the home market. Thus we secure continuous equality of supply and demand. In Switzerland and Norway such systems are at work in the grain trade, and appear satisfactory.

Innumerable examples can be found of new proposals to regulate trade, many of which arise from the special difficulties that have followed over-production of agricultural produce. This over-production and consequent fall in the price of grain brought the problem to the front, not only in the United States of America, in the Argentine and our colonies, but also in Europe, in the Baltic States, in Poland, in Hungary and in the Balkans. In these European states the position so reacted on the peasantry that their debts were said in 1931 to amount to £300,000,000. (69) As an outcome of this situation suggestions were at that time being made in all these countries to regulate production and distribution: even in the backward Balkans, the state was tending to make itself more and more responsible for the organisation of agricultural production. (70) In Australia a scheme has been put forward, which may at the time of writing be in course of adoption, for regulating the wheat trade. In the United States of America a Federal Board, with a fund of £100,000,000 at its disposal and with wide powers of control, has been trying to create order out of the chaos that has arisen in the production and marketing of food as a result of free trade within a tariff boundary. In Egypt, the country of government so fantastic that it reminds one of an Arabian Nights dream, a complete scheme for organising distribution and stabilising prices of food and other products has been suggested by the Government. Even in this country legislative action has been taken since the War to safeguard permanently or temporarily the home market for corn, sugar beet and various industries; further, the Labour Party, whilst in power in 1931, secured the adoption by Parliament of an Agricultural Marketing Act for dealing with the distribution of food products, whilst the National Government in 1931–2 imposed tariffs, prohibitive and otherwise, on many imported articles and also adopted what is called the Quota system for regulating not only the importation of wheat, but also the price to be paid to British farmers.

Countries are indeed nowadays constantly discussing, and even sometimes making, new experiments, principally in relation to food supplies, in tariffs and quota systems, either singly or in combination, in marketing boards, in control and prohibition of imports and also for encouraging exports. Concurrently, proposals are being discussed at Geneva for stabilising the price of wheat in Europe. It is important to watch the effect of all these new schemes.

In addition to these national and international proposals, organisations have been and are being formed within the various states for dealing with home and export trade. A well-known case is the business federation of a co-operative character that is largely responsible for the marketing of Danish bacon for export. It is important to study this federation, of which so much has been said and so little is known since, though it is a sound business proposition, in some of its details it is a clear example of what we want to avoid. This federation was started by farmers with the support of the banks, to build up an entirely new export trade in bacon; it had therefore no established distributive system to compete against. The federation takes responsibility for all processes of curing, transporting and marketing between Danish farmers and the English shops, and to some extent the buyers of other nations. In London it is represented, as already explained, by a Committee that fixes sale prices to the shops, and pays to the farmers of Denmark a uniform price, which is said to be 75 per cent. of this wholesale price. In some respects the effect of this business federation is disastrous, for it takes advantage of the British open market to undermine the British bacon trade, whilst it diverts the Danish farmers from their true business of feeding their own people, who suffer accordingly. This export of food appears to be one of the causes of the relatively low standard of life in Denmark. It may ultimately lead to a serious dislocation in that country, for it is unlikely that Britain will always allow both its agricultural population and its national life to be damaged in this manner. If this country cannot learn from its own experience, it may take a lesson from Canada, where in 1931 the government refused to allow the importation of Danish bacon on any terms. (71)

The reputed success of this and other Danish organisations served to encourage throughout the world other co-operative experiments amongst producers, not dissimilar in character; but such experiments have never, so far as is known, been entirely successful, and many have been accompanied by disaster. This want of success

is no doubt in many cases due to the fact that would-
be co-operators have always had to face the competition
of established distributors, thus creating a trade war
between co-operators and distributors, in which the latter
generally won, and partly because the organisations could
never secure a uniform economic price. All such experiments are bound to fail unless they are based on the
utilisation of present distributors if such be established,
and the securing of a steady price as between producers
and consumers. The greater number of these experiments
both national and international, deal with food products,
but in many countries there are organisations for dealing
with coal and manufactured goods. There are said to
be many such organisations in Germany, whilst in this
country the fixing of a scale of prices, either permanently
or temporarily, to both producers and consumers has
become a widespread movement in many branches of
industry. How extensive it is we cannot say, but this
much we do know: that when the Central Committee
under the Profiteering Acts made enquiries in 1921, and
at their instance 1,853 trade associations were circularised
on the subject, 115 replied stating that they did not control prices, the large majority did not reply, but 446
admitted that they exercised a control. It may perhaps
be assumed that a very large proportion of those who
gave no reply did exercise some control over prices; and
in some cases they appear to have exercised control over
production with a view to equalise supply and demand.
These and innumerable other examples, such as national
and international federations and cartels, can be given of
the present tendency to escape from the system of com-

petitive trade, with the vacillation of prices and irregularity of supply and demand that go with it. This drift towards regulation of trade is indeed one of the most remarkable features of modern civilisation.

The most obvious example of complete reaction against competitive trade is, of course, the postal systems of the world which are uniformly under state control. Here the reaction may well have gone too far, and in this country this possibility is beginning to be recognised. There is a definite turning back in opinion, and the Post Office as a state organisation is being condemned. It is suggested that our postal system should cease to be a department of the state and should be run as a self-supporting business, with special powers and yet subject to a limited amount of state control. Semi-public bodies of the type so suggested have been employed in creating housing colonies and for other minor purposes in this and other countries, and are now beginning to be widely discussed in this country under the title of Public Utility Societies. (72) Such societies, which have long been recommended by sociologists, may be found to be the form of business organisation best adapted for regulating the distribution of food and coal, for building working-class dwellings and for many other purposes. They may indeed to some extent replace limited liability companies in the business life of this country.

With all these experiments before us, and also those that have created the great co-operative movements, (73) we sometimes wonder why so little is done in this country, where the whole problem has been extensively examined and analysed and many practical schemes have been put forward. It is the English passion for a vague idealism that blocks the way, for it is this characteristic that has made us indifferent to practical proposals and peculiarly susceptible to stunts and slogans.

6. OF STUNTS, SLOGANS AND ILLUSIONS AND OF DEMOCRACY

We have now, before we turn to constructive policy, to deal with that great obstacle to clear thinking, the modern passion for stunts, slogans and intellectual illusions. It is hardly to be hoped that we can suddenly convert this passion into an interest in truth and reason, but something may be said on the subject that may undermine the strange belief in their value and efficacy.

Stunts, slogans and the creation of mass illusions are the methods that politicians and many other men and women in public life employ, in maybe perfect good faith, as a help towards the building up of their careers. With their help you may become a recognised leader of the people; you may even become a Prime Minister. But when you have become Prime Minister the first thing you find is that slogans do not provide a solution to the social and economic problems that civilisation has provided for us. That is where every government fails. Neither "Socialism" nor "Safety First"—the two slogans that Prime Ministers put forward in the years before the present national disaster, (74) were found to be of any use, nor does any other slogan help us when we come face to face with the problems.

Let us try and analyse particular stunts, slogans and

illusions that block the way to a clear constructive policy. There is first the belief in the trade war, camouflaged under the words free trade. It has made a special appeal to the Englishman's love of liberty. Then there is "pessimism" and "optimism," and finally "internationalism." Let us take them in this order. We have already analysed the destructive practical effects of free trade in our national life, but the intellectual effects are even more serious. For free trade is based, it will be remembered, on a belief in "the Law of Nature," which itself involves what we call the *laissez-faire* theory. Leave things alone and they will right themselves. The preaching of that doctrine for a century has undermined our capacity for constructive thinking and for national action; at the same time we fail to realise that the power is in our own hands, and that it is for us to reconstruct our civilisation to meet the conditions of modern life. Next in order come the twin illusions of pessimism and optimism, which we cultivate to save ourselves the trouble of thinking. Neither the pessimist nor the optimist troubles himself about the actual detailed facts; the pessimist, viewing the present position, tells us that civilisations always decay, and that ours will do so automatically; the optimist tells us that all will be well—this is only an economic blizzard, it will blow over. "Be of good cheer," said a famous optimistic leader of public life at the moment when chaos was beginning to overwhelm us, and he left it at that. It is unlikely that either optimist or pessimist will be influenced, though other people may be, if it is pointed out that this civilisation differs from all its predecessors in that it has before it a fairly complete and in fact priceless knowledge of the essential facts governing the life of nations, drawn from history, economics and sociology. This knowledge, combined with a new power of analysis and a possibility of concerted action, creating what I have ventured to call "communal free will," three elements substantially new to the world, make it possible to think out the causes and remedies of our present difficulties, and build up, if we can only think constructively, a constructive policy. It is only by taking advantage of these elements that civilisation can make good. To the optimist I might also point out that the present conditions are not the result of an economic blizzard—we have created them ourselves; they arise from a series of definite incidents in civilisation, the outcome of a mistaken policy; indeed, the effect of this policy has been long since so carefully analysed that the present economic state of this country was predicted almost in the exact form that it has developed. We know then that, though there may be temporary improvement of our present condition, this condition is not a passing incident: it marks the end of a period in civilisation—the collapse for this country of an industrial period depending on overseas trade, which was in its turn based on the need of other countries for equipment. Clear your mind, then, of optimism and pessimism, and get down to the facts.

But perhaps the intellectual illusion most dangerous to civilisation is that which is concealed by the word internationalism. "Think internationally" is the slogan that has paralysed the thought of Britain. For the word internationalism, with its six almost rhythmic syllables, has an amazing and to many people an almost irresistible appeal. It has a very special sentimental claim on the mind of the English, who are fundamentally sentimental. It is like a fairy tale: it takes one out of oneself and destroys all sense of proportion. "I think it desirable to treat the world as a whole," was the snub dealt to me on a recent social occasion by a charming provincial lady, a member of the local branch of the League of Nations Union, and so with a reputation for being advanced. I had ventured, in the course of a conversation about unemployment, to make a critical remark about internationalism. I was sat upon; but what did she herself think she meant? I knew that in fact she really thought in terms of her own village; it was all a pose, and so it is to many.

The word internationalism is all the more attractive because there is something in the underlying idea. To be sufficiently cosmopolitan to be able to cultivate a real personal understanding with people of other nations is extremely valuable; cosmopolitanism has also its importance in all that concerns art, literature and music. It is, there can be no doubt, essential to think internationally in dealing with international politics, in order to avoid the political rivalry that, arising between nations, leads to wars and the rumours of wars; internationalism in this sense leads to peace. It appears important that we should have an international standard of money, and an element of internationalism in finance may have value in other directions.

But internationalism in trade and in industry—economic internationalism, as the words have been under-

stood in this country in the past, and are still to a large
extent understood to-day, (75) is in its effect something
quite different. It leads not to peace, but to war: for it
means free competition, it means the trade war, and
its ultimate result, if adopted *in toto*, would be to destroy
civilisation so far as it depends on economic considerations, for it would transfer economic power to the ex-
ploiters of oriental and of backward nations, and so undermine first the industries and then the life of the more
advanced. The very effort to think internationally in these
economic matters seems also to destroy common sense
and develop a megalomania which easily becomes an
hysteria. Overlooking the fact that the world contains
many entirely different forms of civilisation that have
no relation to one another and cannot possibly, maybe
for a century, be brought under common control, internationalists try and think of these civilisations as a whole,
and then wander off to the fantastic idea that everything
so interlocks that no civilisation can be advanced without the others. This is the illusion, and this, I think, is
the reply. The problem of civilisation is a problem of
reconstruction, and unless one is a divinity or a "super-
man," whatever that may be, there is only one way of
dealing with a constructive problem. That is to bring
it definitely within limits over which control can be
exercised. This may not be clear to my readers, if laid
down as a principle, for the English mind does not easily
accept principles. A parable may help. A farmer can only
cultivate a farm by giving continuous and careful constructive thought to the work in all its details. If the farm
is so large that he cannot control it, he will fail, partially

or completely. It is no use for him to think constantly
in large and generous terms of the world at large, or even
of the country in which he lives, for that diverts his mind
from his work and his mind is limited; he has to concentrate on a particular area and make that as perfect
as possible; he will do well to make his farm a model
for his neighbours. In his work he is really limited by
the power of his mind. Every successful constructive
enterprise in business and the world at large is governed
by the same philosophic rule. Britain cannot escape from
it. We cannot deal effectively with matters outside our
control. Though we have to deal with problems in
relation to their surroundings, we have also to isolate
them; if we do not do so we must fail in the future, as
we have failed in the past. For this is a law of life and
we cannot defeat it. It is, for example, a pure illusion to
suggest, as is sometimes done in Labour circles, that we
must concentrate on raising the standard of life in
other countries, a matter entirely beyond our control,
before we deal with the standard of life in our own
country, a matter which lies within our own control.
These are the suggestions of sadists, who want all to
suffer together. (76)

The constantly changing governments of Britain can
each in their turn try and secure themselves from criticism by advising us "to think internationally," but this
slogan leads us into strange conclusions. Let us take as
an example a theory that has recently become popular
amongst economists and politicians of all parties. They
say, in various words, that we cannot deal effectively
with the problem of unemployment in this country

because it is a world-wide condition. A doctor might just as well say that he could not deal with a patient's stomach ache because a stomach ache is a world-wide complaint. The statements of the politicians and the doctor are equally unreal, for in both causes the disease is world-wide, but the remedy is local. In fact, the world-wide prevalence of unemployment makes it peculiarly easy to diagnose the disease and helps us to indicate the local remedies. Even an internationalist could see this if he took the trouble to ascertain the facts; but that is the trouble with the internationalist, he protects himself by a word to which he gives a spiritual value.

So long as men and women are free traders, optimists, pessimists or internationalists, their minds are paralysed, and they cannot see the truth. The first thing is to clear our minds of these particular "isms," and the stunts and slogans by which they are supported. Solutions emerge from knowledge of the facts.

Many of our most distinguished thinkers on political, sociological, financial and economic problems seem to be carried away by illusions. I am inclined to think that this has always been so. Such men and women are unusually sensitive and seem easily affected by popular opinions—by mass suggestion. They live in a world of their own, remote from the hard facts of life, and ideas from the outer world penetrating into their atmosphere are surprisingly often accepted without analysis. They are then woven into their thoughts and arguments. Their conclusions are accepted by the public as if they were the Ten Commandments when they are really much more of the character of the first chapter of Genesis. It is certainly astonishing how little some experts of special aspects of social problems know of the surrounding incidents. As a general result the most unreal ideas are even to-day constantly found in the writings and wireless pronouncements of distinguished individuals; very often, too, so disguised by skilful phrasing as to be easily accepted without criticism. This mental attitude may account for a book like Adam Smith's *Wealth of Nations*, with its curious jumble of truth and falsehood, and for many other later works. It may also help to explain the views of some at least of our modern thinkers. The power of obsessions in economics and sociology is all the more curious since in other matters reason has already replaced imagination. Astronomy has long ago taken the place of Astrology, Chemistry of Alchemy; we no longer believe that the world was made in seven days. In all these things we examine the actual facts and reason from them. But in dealing with our social problems we are still far too much governed by imagination; it is thus that we overlook the truth. Most of us have yet to learn a simple fact that a distinguished University professor, a few years ago, blurted out to a girl student of Economics. She had asked him with feminine directness whether his teaching of the theories of the orthodox English Economists was true. "Certainly not," he said. "Nothing really happens like that; economics is only an examination subject." Economics is often taught not as a science but as a dogma to be accepted as an "act of faith." (77)

We can only hope that the spirit of reason which controls thought in the world of science will rapidly spread its influence to economics and sociology; there are signs that it is beginning to do so. Meanwhile we have in public life to face the problem of bringing truth into an atmosphere of illusion. It can hardly be presented in a crude and indecent nudity. It may have to be clothed with illusions and decorated with slogans. One can only hope that this will not be necessary.

It has often been assumed that democracies would be at the mercy of "stunts, slogans and illusions." I very much doubt it; it may be found that illusions are really an intellectual disease. They were certainly more powerful in the pre-democratic XIXth century than they are to-day. The free trade illusion, for example, carried away the thinkers of the last century, but now that democracy is established it is rapidly becoming suspect. Even that attractive slogan "Empire Free Trade" (78) with all the money and influence that lies behind it, has little effect. "Socialism," belief in which was at one time almost a religion, shows signs of losing its power to-day. That strange devotion and belief in political leaders who were elevated almost to the position of demi-gods in the last century is certainly disappearing. The democracy takes a quite unexpected interest in facts and practical proposals. A Conservative leader, accustomed to deal in facts, remarked to me one day after the election of 1929: "This is the first election at which my audience really wanted to hear what I had to say. They seem to want to know the facts." I believe that is to-day a common experience; it has certainly been my own. An interesting result, entirely unsuspected by politicians, may follow from the wide extension of the suffrage. Although the election of 1931 does not perhaps lend support to the idea, the electors may become impervious to stunts, to slogans and to illusions, and many turn when perhaps it is too late to facts and reason.

PART IV

REBUILDING BRITAIN

PART IV

REBUILDING BRITAIN

I. INTRODUCTORY

We have now to consider the steps we have to take to-day in this country.

And here we come right up against the special obstacles that have prevented action for generations.

The form our civilisation has taken has gone far to destroy the power of constructive thought. It should be intuitive, and in some persons it is still intuitive; in others it appears to be entirely lost, or else to exist in a rudimentary form, only to be employed with difficulty and often with real mental suffering. We have now to cultivate constructive thinking as an art; it is essential in the solution of social problems. Analyses, such as those given in the earlier part of this book, are only of very limited value, for the simple reason that the future does not evolve out of the past, but has to be created.

This want of power to think constructively has an even more serious result, for it not only prevents individuals from building up constructive schemes, but also forms an obstacle to their being understood when presented for consideration.

Notwithstanding all these difficulties, it is well to face the facts and submit constructive proposals; even if they are not in all details correct, they do excite imagination, and may lead to action.

L

When, then, we come to consider what exactly we ought to do in this country to-day, we find two conflicting points of view; with these two points of view go corresponding lines of policy. It is well to understand both points of view and both policies, for though they are held by relatively few persons, they profoundly influence the thought of the great majority of the nation, who are too much occupied with the details and difficulties of their own lives to give much consideration to national problems.

There are those, a little discredited to-day, who, pinning their faith maybe on the phrase "The Economic Blizzard," the phrase popular during the years of the Labour Government, look on the present world condition as a storm that will blow over. They observe the affairs of the world in very much the same spirit as we contemplate the weather: a matter over which we have no control. Convinced that "the Blizzard" or "the Crisis"* will pass, they are prepared to be satisfied, to take a phrase from the lips of the present Prime Minister, Mr. Ramsay MacDonald, with "a series of piecemeal and ephemeral compromises of a purely temporary character." Thinkers of this school inspired the Labour Government of 1929–31 to carry out a policy of laying in a stock of what they call "emergency measures," such as road-making and other public works, to provide occupation until times change; they may have influenced the early policy of the National Government in putting forward in November and December 1931 temporary measures of control of trade which were supplemented by the

* With the coming in of the National Government the "Economic Blizzard" was rechristened "The World Crisis."

valuable but somewhat sentimental appeal to the nation to "Buy British." (79) With this goes a policy of protection against the storm; doles from the insurance fund, poor law relief, pensions and so forth. The most logical of these protective proposals was that put forward by the General Council of the Trades Union Congress in their evidence before the Royal Commission on Unemployment Insurance in 1931. (80) It was their suggestion that every employed person should make his contribution to a special fund that should provide every unemployed person with an income until work was found for them. Theirs is a policy of macintoshes, umbrellas, gaiters and even goloshes—complete protection against the rain. The victims of the storm are no doubt to console themselves from time to time by studying the weather forecasts of the political weather prophets. There will be always some amongst the politicians in power who will see signs of a trade revival and tell us "to be of good cheer." There will always be others, out of power, who will be prepared to predict the passing of the storm, if, and when, they assume the reins of government. From either comes the hope that springs eternal in the human breast and protects the mind from the trouble of thinking —a laborious and wearisome business for which the human mind is ill-constructed!

Then there is another group of thinkers with an entirely different point of view. They hold that we are at the end, or nearly the end, of a special period of civilisation, that may well be called the period of World Equipment. They point out that this period of equipment did, in fact, owing to the accident that this country was first in the field, create for Britain an industrial boom. That, they say, is over. Britain, at any rate, is clearly at the end of her industrial period. It arose out of special circumstances that cannot be expected to return. They also point out that even if there is some revival of overseas trade—the outstanding feature of the past—such revival is likely to take one of two forms, either the further equipping of nations to provide for themselves the requirements that in the past Britain has produced for them or else the export of goods that will be paid for in food and manufactured articles that we could well produce ourselves. The ultimate result in both cases may be expected to be increase of unemployment. Moreover, they believe that the metaphor of the weather, with talk about the Economic Blizzard, confuses the issue. They accept neither the "Law of Nature" nor the belief in predestination that goes with it. Our fate, they say, is in our own hands, and we have therefore to adopt new methods to build up the industrial side of our national life. We have to make our own history. They are, therefore, definitely in favour of planning a constructive policy suited to the new conditions, and of taking the steps needed to put it into force. To do this, they ask the nation to dismiss from its mind belief in abstract theories and accept and work on the basis of the facts of the past and present. They say, further, that the problem is a problem of reconstruction, and that an essential element in reconstructive work is to limit it to a particular unit over which control can be exercised, and to put the direction of the work into the hands of people who have both the interest and the power to carry it through. If Britain, they argue, is to do anything effective to reconstruct the economic side of its civilisation, it is necessary to recognise that its constructive work can only be planned and carried out within such a definite framework, that is to say, the framework of the nation, for it is only within the national framework that there is the power of regulation of trade and finance that makes action possible. They are, therefore, in this respect opposed to internationalism: they are nationalists because, realising the limitations of life, they want to get something done.

Between these two schools of thought there is really no compromise. Nevertheless, so strong is the belief in compromise based on the habit of evading facts, that innumerable intermediate and small remedies have been and are being put forward. It was John Stuart Mill who wrote, "Small remedies do not produce small results, they produce no results," and he is substantially right.

If, then, we are not prepared to "wait till the clouds roll by," we should be planning out a large, bold, comprehensive and genuinely organic reform in our industrial life,* and creating a driving force to carry it through. This work of planning constructive schemes faces all the civilised nations of the world to-day. It should have a special attraction to the adventurous Briton, for it is a real adventure in civilisation; indeed, for a democracy to plan its future and to carry out its plan is something entirely new in world history. In the past dictators may have thought out and carried through constructive policies. In relatively modern times, some

* The policy should obviously be directed to the absorption of the whole of the unemployed into productive work.

three hundred years ago, the dictators of Japan appear
to have made a definite plan and carried it through;
they decided to cut themselves off from the world and
build up a national civilisation. The result was the creation of a national life which, strange as it seems to us in
many of its details, was nevertheless of extraordinary
interest and beauty. When, in the latter half of the last
century, the pursuance of this policy was made impossible
by the aggression of other nations who imposed their
will on defenceless Japan, the position was faced and
grasped. Again a plan was thought out; it was decided to
give up their policy of isolation, and at the same time to
adapt the industrial system to their national life. In
both cases it was the intellectual power to make a plan
and the ability to carry it through that were the striking
features—but the people were not consulted; it was done
by what was substantially a dictatorship. Something has
also been done since the war, and is being done to-day
from the top. In Czechoslovakia, immediately after the
war, Alois Rašín introduced and imposed upon that
nation a new financial system, of which too little is known
in this country; it appears to have had great value in the
reconstruction of the finances of that state and in development of its industry. (81) Mussolini has obviously a
constructive plan for the national growth that he is building up in Italy. The little group of dictators who have
managed to secure the control of Russia have a plan which,
with the help of a small though highly organised section
of the community, they are endeavouring, apparently
with some measure of success, to impose upon the
various races that inhabit the immense areas in Europe

and Asia that come under their control. Their policy is directed to mechanising life and industry. (82) It seems fantastic in some of its details, and it appears in certain cases to be enforced with the cruelty that has always been the characteristic of Russian governments, but it is at least a plan. All these plans are being carried out by the force of what is substantially a dictatorship. Even the minor reconstructions carried out in Central Europe after the War, in which I took a small part, were directed from the top, and were not understood by the people themselves. The Polish Governor of the province made the position clear to me. "The peasants," he said, "of my province have no idea who you are or why you are doing this work; but they are prepared to do what you tell them." Then he added: "You must not be depressed; in two generations you will be a saint, and they will take an annual holiday in your honour." It is only in Switzerland, Holland and the Scandinavian countries that we find new and definite industrial policies that have been planned and adopted by democratic nations.

The British problem is therefore an adventure, for no democracy has ever moulded its own future. It is this that should give it the real appeal to the British people.

It is perhaps fortunate for the chances of success of this adventure that the present chaotic conditions in the economic life of the world have long since been foreseen by certain sociologists and economists who have created what is substantially a new school of constructive thought, concerned in dealing with the problem of reconstruction, of which the late Lord Milner was the most distinguished figure. (83) By this school (on whose analysis the proposals here put forward are based) the various aspects of the modern problem have been analysed and solutions thought out. Their work has been done in advance and with care. But it has received, until recently, far less attention than it deserved. To most of our leaders, indeed, the problems of civilisation did not seem sufficiently acute to make it even necessary to consider the fundamental ideas on which that civilisation was based. Definite constructive proposals, however valuable, were as a rule smiled upon as effusions of amiable and simple-minded cranks. It was only in the year 1929, when the rapid increase of unemployment indicated to many thinking people that the end of an industrial period was approaching, that an atmosphere favourable to new constructive ideas began to be created. Since that time, newly popularised ideas, arising in a surprising number of cases from the teaching of the modern school of thought referred to above, have been considered by politicians and the press. There has even been some measure of acceptance, with the result that such new ideas, hurriedly absorbed from the teaching of the modern school, but often I fear insufficiently understood, appear in many politicians' speeches, in press articles and even in legislation.

We are constantly told, often with immense emphasis, that we must "mobilise our economic forces"; but there our advisers stop short. They do not tell us what is to be done with our economic forces when they are mobilised. I have already suggested that the primary object of national policy should be to provide for the population of this country the four main essentials of life—food, houses, clothing and lighting and heating—supplementing this provision with the development of the amenities and pleasures of life which may be grouped as health, knowledge and happiness. It is with this object, I suggest, that we have to "mobilise our economic forces," that is to say, we must absorb our surplus energy represented by the unemployed and direct it to this special purpose with all the skill and constructive ability that we can gather together. Further, in working for the attainment of this object, we must utilise scientific knowledge and our national resources, and also (so far as they do not conflict with the interests of Society) the powers that come to us from machinery.

It is not possible in this essay, which deals in the main with one aspect of our social problem, the question of trade, to give complete proposals for reconstructing our national economic life in accordance with this line of policy, and to support these proposals with the mass of illustrations and arguments that would make the case absolutely convincing. I therefore propose merely to indicate what I believe should be done at once towards developing our natural resources, and to give in some detail the special case for agricultural reconstruction, a problem which has been fully worked out.

2. A LINE OF POLICY

(a) INDUSTRIAL DEVELOPMENT

The first point that we have to recognise is that this country has drifted into a position where we are primarily concerned in industry. The mass of the workers of this country are trained for industrial work, and factories are standing empty or half-empty, ready to carry it on. We cannot evade these facts; and we have therefore to see how far industry can be restored.

It is quite clear that there is little hope for any permanent revival of overseas trade, but it will be remembered that at the present time we are importing for the home market manufactured goods of a value of between £200,000,000 and £300,000,000 a year that we can produce ourselves, whilst our workpeople who might be engaged in this work are standing idle. The cost of this idleness is enormous.

We have clearly, then, so to arrange our economic life that so long as we have unemployed workers our manufacturers have the first claim on the home market at a fair price. The immediate steps that should be taken to secure this result appear to be the imposition of import duties on manufactured goods we can produce ourselves. This is not by any means an ideal remedy; control of imports is better, for tariffs stand for punishment, and control of imports for prevention of what is really an economic crime. However, it is immediately applicable; it may have to be supplemented by some control of prices, and it should ultimately lead to a more scientific method of controlling imports and regulating internal trade on the lines that are indicated later for agriculture.

But this restoration of a section of our industries can hardly absorb more than a million of our unemployed, whilst there will be, as has already been pointed out, some, though not great, loss of employment in the shipping industry. With this development must go, as the second string to our bow, the revival of agriculture.

(b) AGRICULTURAL REVIVAL

I am not myself a believer in what has been called an "agricultural civilisation." I take the view that common sense comes from the country, civilisation from the cities. What is wanted is to secure the "equilibrium of Society" by maintaining a balance between the basic groups of industries that provide the material essentials of civilisations—food, housing, clothing, and lighting and heating. In this country it is agriculture that is under-developed, and is, therefore, open to development. It is unnecessary to emphasise this point, for the importance of reviving agriculture is obvious to persons whose minds are not warped by inherent prejudices and illusions, and in dealing with it I am only saying once more what has been stated in the last quarter of a century by thousands of intelligent thinkers of both this and other countries.

Nevertheless, since we are a democracy, the case for agricultural revival has to be restated over and over again, year after year, until it is understood by the people and accepted by the politicians.

From every point of view the agricultural question is to-day the central problem.

We may consider it first in its relation to unemployment. The revival of agriculture naturally suggests itself to anyone familiar with the history and origin of unemployment—it is indeed not a bad plan to trace things to their origin. Unemployment has in the past always been born in the villages, though it has grown up in the towns.

In the last hundred years about a million workers and
their families have been dismissed or have withdrawn from
agriculture. Many went to the colonies, but large numbers
always went to the towns and industrial centres, to squeeze
out the weaker city workers. Even in the worst times
there has, until quite recently, almost always been a
job in the town for a rosy-faced man. Such rosy-faced
migrants have often described to me how they have been
picked out of queues of unemployed waiting for a job.
Further, in recent years the younger men have been
constantly drifting out of agriculture to take up special
work. When, for example, a few years ago Hastings was
enlarging its esplanade, men came in from the countryside to work as labourers, and when the job was done
they were no longer technically "agricultural labourers"
but "general labourers," so they went, as they said,
"on the dole." Similar stories come from all over the
country.

The million of the rural exodus was roughly represented
before the beginning of the present chaos, in the year
1929, by the million at that time permanently unemployed.
This figure of a million, curiously enough, comes up again,
for there is at the lowest estimate employment for a
million workers in agriculture and relative industries
to-day.

It is extremely important to see clearly the relation of
the unemployment problem to agriculture. It can be
stated broadly thus. If food that we might produce in this
country comes in from other countries as "tribute trade"
to pay interest and dividends on foreign investments,
or to pay reparations, there is, as has been explained

before, no return in coal or manufactured goods; not only agricultural, but also some at least of our industrial workers, who might be providing manufactured goods to meet the requirements of the agriculturalists, are thus kept out of employ. If this food comes in as "barter trade," there is a return of coal or manufactured goods: then one set of workers only, the agriculturalists, are kept out of employ. But if we produce this food ourselves, and incidentally create a home market for industrial workers, there is work both for agriculturalists and industrialists.

It will perhaps help to make this clearer if we take approximate figures. Our home output of food products was, in 1931, computed in round figures, at the wholesale prices then prevailing, at about £200,000,000 a year, and our requirements were about £560,000,000 a year. It seems generally admitted that we could with advantage double our food production in this country. There is then at least £200,000,000 worth of food coming into this country that we might with advantage produce at home. If all of this, to make a theoretical analysis, were to come in as "tribute trade," it would keep out of employ about a million agriculturalists and a very large number of industrialists; how many it is impossible to estimate —half a million is perhaps a fair estimate. The cost in unemployment benefit for a million and a half workers would in that case be about £75,000,000 a year. If this food comes in as "barter trade" there is corresponding employment for industrialists in providing the coal and manufactured goods sent in return, but the importation still puts the million agriculturalists out of employ, and the cost to the nation is £50,000,000 a year. If we produce the food ourselves, the million workers are employed in agriculture with related employment in industry, and the cost of unemployment disappears. The loss of employment in the shipping trade has to be considered: relatively it is not great. Indeed, foreign countries often insist that their trade shall be carried over the seas in foreign bottoms. We can put all this from a slightly different point of view. I suggest that to arrive at the true cost to the nation of food coming into this country which we might advantageously produce ourselves, we have to add from 20 to 40 per cent. to the price paid, an overhead charge to pay the cost to the nation of the related unemployment. (84)

We have also to consider the national and personal degradation that unemployment causes to the people of this country, and if we are cosmopolitan and have sympathies with the actual workers, we may, for example, remember the tragedy that has come to the people of those countries who, having been enticed by possibilities of our open market and drawn into the trade war, have attempted to supply us with wheat and other food, with disastrous results to themselves. If we are only imperialists and yet have sympathies with the workers, we may yet consider what has happened, for example, to the farmers of Canada, whose plight in recent years, a reflex effect of our open market for wheat, has admittedly been deplorable.

There is still something more to be said of the relation of agricultural revival to the problem of unemployment. A large part of the reconstruction work that will have to be undertaken in rural districts, if there is to be an agricultural revival, is of the nature of equipment: the erection of farm buildings and cottages, the institution of jam, bacon and other factories, the provision of electricity, of drainage and also of a requirement constantly overlooked but of enormous importance—irrigation and water-supply. This equipment will give wealth-producing work to men now unemployed who are industrialists by training. Then there is to be considered the purely agricultural work, to be carried on when the arrangements are completed; but much of this has been mechanised, tractors and other forms of machinery are already in widespread use. There is also the employment in the factories for converting produce into food and many minor supplementary trades. The importance of afforestation in this connection must not be overlooked, and here it is noteworthy that in this agricultural and associated work a large part does not require workers of a special agricultural bent. In fact, an agricultural revival provides employment for many years for industrialists in the work of equipment and also permanent employment for men and women of all classes; and there are men and women of all classes who need work. There is no necessity, it should be observed, for the large majority of these workers to live on the land; indeed, with certain definite exceptions, there is no more need to-day for a land-worker to live on the land than for a stock-broker to live over the Stock Exchange. Land-workers can live in villages, where life, it may be noted, is no longer dull, in country towns, or in new towns built in the centre of agricultural areas. Finally, the revival of agriculture will provide the best possible home market for the industrial workers.

So much for the effect of an agricultural revival on unemployment. It is also important to remember that this revival would provide not only employment, but the increased production of national wealth required to enrich our people and to maintain and raise the standard of the life, the essential of a developing civilisation, and also to reduce the burden of taxation by spreading it over the increased wealth so made available.

There are other points to be considered that favour agricultural revival to-day.

Britain is a country peculiarly suited for our purpose. Provided that we cultivate to its full capacity the large areas that are only half-developed, bring into cultivation the small areas of waste land now neglected, install electric equipment, drainage and irrigation and generally employ modern methods, there can be no doubt that we shall be able to produce the greater part and possibly the whole of our essential food supplies. Moreover, we have a moist, temperate climate, (85) whilst the soil, though extraordinarily varied, is in many cases remarkably good, and in others open to improvement with the help of fertilisers and more scientific treatment. Then there is the remarkable character of our cattle and other stock, or at least the best of it, eagerly sought for breeding purposes by the farmers of all the more progressive countries in the world. We have also admirable seed supplies, especially of wheat, heavy croppers such as Square Head Master and Revil Wheat, and a more recent addition, Yeoman II, which has all the qualities of the best mixtures that English millers and bakers use. Further, we must note the special genius of our leading farmers, and the alertness and keenness of the great majority, especially of the younger men. We have also available a large actual and prospective population of agricultural workers suited to be either smallholders or labourers. It is indeed not only the men still remaining in agriculture who have agricultural capacity, but the towns and industrial centres are full of men and women of agricultural stock, many of whom would gladly return to the countryside if rural conditions were made sufficiently attractive; whilst the remarkable success of the allotments set up during the War and more recently founded in connection with the movement for finding work for the unemployed show that even amongst industrial workers there are very large numbers of men specially adapted by temperament for agricultural work.

These special features are supplemented to-day by the priceless experiments being made by our many distinguished scientists. The results of these experiments are being disseminated by practical men employed by County Councils and other authorities, who are giving, within the areas of their activities, invaluable practical advice which is eagerly absorbed by a considerable proportion of the working cultivators.

There are many other important points, of which the one most often overlooked is that home production of food will give permanent protection for the British consumer against high prices; high prices are a real danger, for it is important to realise that if the present system of purchases from abroad continues, in a very few years prices may go up to an artificial figure, as a result either of the reduction of production abroad or the definite action of cartels or groups organised outside the country. Indeed, it may well occur, in a very few years, that in the essential matter of our food supplies, Britain may become entirely at the mercy of foreign financiers and traders. We may be robbed or starved. Moreover, such reconstruction will be the first, though only the first, step—for there are many other things to be done—towards creating the balance between agriculture and the three other great services, housing, clothing, lighting and heating, on which our economic life depends. It will also make it possible to provide the British people not only with food, but with fresh and wholesome food; to decentralise our population and so help to solve the problem of the slums; to give to the people at large a share both in the civilisation that comes from the towns and the common sense that comes from the country. It will, in fact, help to create a healthy, intelligent nation.

To all these facts but one reply is given and that is a theory: it takes the form of a slogan: "Your food will cost you more." This slogan has been repeated by politicians for generations, and for that reason is widely believed. But why should home-grown food cost us more? It is much more likely to cost the nation less. For even putting aside the cost of related unemployment, which adds, as has been pointed out, from 20 to 40 per cent. to the wholesale price of imported food, it has to be realised that the actual cost of food production in this country is singularly low; it is indeed not the cost of production, but the system of competitive trade reflected in costly distribution, that causes home-grown food to be dear food. There is no reason to suppose that even under present conditions, if we give up our belief in our present competitive marketing system and decide to rationalise the distributing trades, that the home-grown food that we should naturally produce in this country would cost us on the average more than the corresponding imported food. There is a wide margin for savings in distribution. Moreover, with the help of increased scientific knowledge and new equipment taking the form of the provision of electricity, drainage, irrigation and water-supply, it may be expected that there would be an even larger margin available not only for well-earned pay and profit for the producers, both farmers and labourers, but even, thanks to what is called "the law of increasing returns," (86) for cheapening food for the consumers.

If we admit that the revival of agriculture should be our first step, it is well to realise quite clearly what our aim is. The agricultural problem is not merely an affair of saving the position of our present farmers and preventing the dismissal of workers that is going on day by day; it is not even as Dr. Addison, the Minister of Agriculture in the Labour Government of 1929–31, supposed, only a matter of increasing our agricultural output by 50 per cent. and perhaps providing employment for half a million men. (87) It is something much more important. It is a problem of making agriculture so attractive as to draw back to the cultivation of the land promptly at least the million workers who have been driven from it, and properly belong to it; whilst we may hope thereafter to go on with this agricultural development in conjunction with afforestation and industrial reconstruction until if it be possible the whole of our present unemployed have been absorbed in useful work.

What now have we to do? It is really something quite simple. We have, so far as possible, to give the farmer security both in his market and his price. It is this second change, the reintroduction of the system of fixed instead of vacillating prices, that will alter the whole basis of our economic system; it leads to the production of wealth instead of the making of money. It is important to understand this quite clearly. When prices in agriculture fall below the economic level it constantly happens, as has been explained above, that whilst the national interest is to increase production and employment, the farmer's personal interest is to reduce production and dismiss his men, for he finds that it is only by taking this step that he can make a living. But when the farmer has a fair price to work to, his self-interest, that is, his desire to secure a living for himself, and the national interest, that is the need to enrich the nation by the creation of wealth, become reconciled. The definite result of working on a system of prices fixed at an economic level is that the more the farmer produces the better he will do for himself and the more he will benefit the nation.

Courageous readers of this essay, especially those who are accustomed to think constructively, will perhaps not be unwilling to hear in some detail how this can be done. We have to adopt a system which is the antithesis of free trade, and I submit one that has been worked out in sufficient detail for immediate political action, and has support from the present Prime Minister, Mr. Ramsay MacDonald. (88) It involves the adoption of three perfectly practical proposals.

I. *National guaranteed standard scales of prices* for at least the main articles of food production. The principles on which these prices should be settled are still little understood, and may therefore be restated here. In considering scales of prices, we have to put aside all the free trader's theories about the value of "world prices" and to limit ourselves to the national interest. We have therefore to weigh the effect of any particular price upon the national problems of: (1) employment and wealth production; (2) raising the standard of life for the workers on the land and on securing a fair margin of profit to farmers, particularly important in order to make agriculture an attractive employment, able to provide funds for agricultural development and to secure from the countryside a good home market for manufactured goods, and (3) the obvious need for securing a fair and if possible a lower price of food to consumers. Economic prices for food will be found to be neither high nor low; they will be medium prices.

II. *The control of imports.* This limitation is required not only to bring the whole problem into administrative control, but also to secure to the British farmer the first claim on the home market at a fair price; it can be carried out immediately by a Marketing Board, as in Norway, or a system of licensing, as is established to-day in Switzerland, under which merchants cannot buy from abroad without a permit. Import duties are not essential or particularly helpful, but a small duty may be desirable as a method of covering costs of administration and other related expenditure.

III. *The scientific organisation of distribution.*—This is a problem not for farmers but for distributors; we have to rationalise distribution—that is, to co-ordinate the work of our distributors in each of the various branches of the numerous trades that go to make up agriculture, with the object of securing that distribution of food products is carried on as effectively and simply as is the work of distributing letters in the Post Office. In the early stages of this work of co-ordination we should be concerned in reorganising the present distributors so that there should be no waste or overlapping.

Knowledge of the problem suggests that the form of organisation best suited to secure these three fundamental reforms would be the formation of what have been called "Distributive Guilds" or non-profit-making "Distributive Trusts" based on the co-operative principle, whose primary function should be to take the British farmer's produce at the standard price, and to arrange for its conversion into food and delivery to the retailer in the quickest and most effective way. Into such organisations the present distributors might well be absorbed. It will be possible when we have perfected such a system to secure savings in the cost of distribution sufficient, if divided between producer and consumer, not only to raise the farmer's wholesale price to the economic level, but to make substantial savings in the retail price. Concurrently, the distributive organisation could secure from abroad the deficit, that is, the food that cannot be produced in this country within the limits of the economic price. This is undoubtedly the ideal form of organisation, but it is unlikely to be adopted until it is better understood; still, it is well to bear it in mind. The immediate problem is to work out the details of schemes of standard prices and distribution, with the help of the present distributors and the converters of raw material into food. (89)

These are the three practical steps that have to be taken to secure the revival of agriculture. It is well to face the fact that in bringing such a new system into working order subsidies may, in some special cases, be necessary. They might be provided out of the funds allocated to relief of the unemployed workers who could be absorbed in the wealth-creating work of food production, or, better still, by a system of small import duties which would incidentally enable us to keep a hand on imports and also provide funds for administrative purposes. But subsidies should be quite unnecessary when the new system was in working order.

(c) Method in Reconstruction

It is of little use to prepare a plan, however sound, without also providing a driving force to carry it out. There is indeed no possibility of real success for any piece of constructive work unless it is in the hands of a capable directing individual or group of individuals, not only keen and concerned with success, but in a position to work continuously at the details and to tackle emergency problems as they arise. We in Britain are far too prone to leave things to luck, and, indeed, in the past we had all the luck. Ever since Elizabethan times we have been going in for wild-cat adventures all over the world, and somehow we attained a measure of success. These adventures, when they succeeded, and it is sometimes forgotten how often they failed, won through because of the special resource and genius of the directing individuals. But in the more serious work of rebuilding our civilisation we cannot rely on the luck of the adventurer, any more than on the chance of muddling through. We must not only eliminate those factors which are outside our control, but establish a control over what is left after the elimination. But such control must leave the constructive workers as free as possible.

We need, then, not only a plan but directors. Every capable business man who has had to build up a new business, or indeed carry on an old one, realises this need; and every woman who has had to run a household, or even to bake an apple-tart, knows it intuitively. But when we come to public life it is constantly overlooked. I hope I shall be forgiven if I emphasise this point, which is of overwhelming importance, by an illustration from my own experience, slight in itself, in the reconstruction work in Central Europe to which I have referred before. I remember very vividly exactly how this problem of direction arose when I was set down with instructions to reconstruct agriculture in the devastated province of Dróhicyn in the White Russian areas annexed by Poland after their little private war with the Bolsheviks in 1920. I started tentatively and hesitatingly consulting the people

on the spot and trying to get their help and approval;
I approached the problem in a democratic spirit. I made
little progress. But one day I picked up, almost by accident, a young Cossack aristocrat, a refugee from Deni-
kin's army. He had a profound knowledge of Russian
peasant character and of the texture of village life, which
in fact, as I quickly discovered, closely resembled the life
of English communities of the XIVth century. He had
also some experience of administration; moreover, he
had, like many cultivated Russians, what was even more
valuable—a philosophy of life.

At first he merely watched and helped in my work;
he obviously did not entirely approve. One day he said
to me rather shyly: "Pan Fordham, may I say what I
think?" I encouraged him to go on. "I think," he said,
"we are working in the wrong way. I understand," he
went on with a rather sardonic smile, "your Western
theories of democracy, but they cannot be applied here.
These people are semi-orientals, half-civilised. They do
not understand you or your method. What is wanted is
direction; you must give orders." We discussed the problem from time to time as we sledged over the snowbound
countryside, for like most Russians he loved talking.
Finally I agreed. I made myself a director; jocularly I
said that I would be Governor and he was to be Prime
Minister. Then we got on. The Polish Governor of the
province was very much amused. "You are more powerful
than I am," he said to me laughingly one day. "If you
want anything done, you can hold up supplies, and I can
only send the police." The police, as it happened, were
incompetent and, I fear, corrupt.

With a directing policy and control we succeeded, and when spring came we soon got a couple of thousand peasants established on the land, with their little strips ploughed and sown. Once started, the constructive process went on, although I myself went back to other work in England.

One cannot, of course, have a director of agricultural reconstruction in this country with the powers that I assumed in that little province in Central Europe. But we must have direction, and its form must be appropriate to our national circumstances. There is a vague belief in this country that the direction of such constructive work is the function of Parliament. We have to go to Parliament to lay down a general line of action, but it is impossible to conceive any group of people less suited than Parliamentarians for carrying out the detail of a constructive policy. They are also far too busy. There is, moreover, no continuity in Parliament—any government is, in fact, liable to change the policy of its predecessor. Now, continuity is the essence of success.

We will now, following the procedure I have taken in reference to the planning of reconstruction, deal with the problem of administration in its special relation to the reconstruction of agriculture.

The absence of any authority with power even to decide on important problems of reconstruction was well illustrated by an incident that occurred following the entrance of the Labour Party into power in 1929. At that time this party had great ideas of doing something for agricultural reconstruction, and a National Agricultural Conference was called. This Conference consisted of the Minister

of Agriculture and his group, representing the government, and therefore presumably national interests, and
a number of individuals selected from the National
Farmers' Union and the Labourers' Trade Unions and
organisations of landlords, to represent agriculture.
Neither group had any real authority to agree terms.
The Minister, on the one hand, was bound by the political
theories of his party, whilst on the other hand those who
were supposed to represent agriculture had no corporate
authority, and according to all accounts differed amongst
themselves. Some of the members of the Conference were,
moreover, politicians, with the passion for tactics that
is the ruling feature of politicians. The talk went on for
weeks and months; nothing happened. It ended in a
stalemate.

But the very failure of the Conference made an impression on the more intelligent of our agricultural
leaders. For the first time they seemed definitely to realise
the practical importance of a point that had been explained
by sociologists and economists for at least a generation
—the need for creating authorities outside the political
sphere, that would be able to decide and to act. They also,
perhaps, began to see more clearly that there were two
interests involved—the interest of the nation and the
interest of the agriculturalists; and that they had to be
reconciled. From the very failure of the Conference
emerged the idea that both these interests have to be
organised and be given power to settle problems and take
action. It was, curiously enough, the Central Chamber
of Agriculture, the association that had been, for some
reason, excluded from the Conference, that first realised

the position and, having done so, promptly took action.
They have as a first step asked the government to constitute permanent authorities to deal with the problems
of agricultural reconstruction. (90) The powers of these
authorities have already been provisionally defined, and
it is clear that they should be sufficiently wide to allow
them at once to prepare the necessary schemes for
reorganising marketing, controlling imports and stabilising prices. The wider the powers entrusted to the
authorities the better; but if we are too timid to recognise
this fact, as an alternative their schemes could, before
adoption, be laid before Parliament for confirmation.
Subject to this confirmation, the authorities should be
empowered to take the necessary steps to initiate and
carry on the schemes. It is certainly of great importance
that they should have wide powers to deal with control
of food imports, so that they can, in case of emergency,
act on their own responsibility, in the prevention of the
dumping of food products from abroad.

This is a simple and practical proposal, and, what is
more important to realise, the only practical proposal
before the country for creating order out of the present
agricultural chaos. Organisations of this character are
the commonplace of government. They were constantly
employed in this country in the past, and have reappeared
in modern times; when, for example, the Small Holdings
Acts were passed in 1907-8, Small Holdings Commissioners were appointed on the initiative of a group to
which I belonged, and proved to be invaluable, whilst the
Minister of Agriculture in the late Labour Government
was persuaded, on my initiative, to introduce into his

Agricultural Marketing Act a provision for the appointment of Marketing Commissioners to aid its administration. Special authorities are also being rapidly created to-day for dealing with problems, such as Electricity, the Coal Trade and Transport, not only in this country but throughout the world. The value and importance of such authorities will be accepted by everyone whose mind is not warped by the illusion that Parliament, which should really concern itself with legislation, ought also to be an administrative body.

By securing a responsible National Agricultural Authority, we shall create a central directing organisation representing the national interests. There remains to be created an organisation representing agricultural interests. There is no such organisation to-day. The National Farmers' Union represents a large section of the farmers. The Agricultural Trade Unions represent a small section of the labourers. But both are essentially trades unions, and the very large majority of the persons connected with agriculture remain outside. There are, however, other organisations, with no very clearly defined powers, that do, or at least might, represent agricultural interests as a whole. The National Council of Agriculture is a statutory body that should be, though it is not, the Parliament of British Agriculture, whilst Agricultural Committees for every county are attached to the County Councils. In addition, the Central and Associated Chambers of Agriculture, an English organisation, and the corresponding Scottish Chamber, are old-established organisations. It is not the place to give an exhaustive scheme for creating a Federation of the agricultural interests, but it is suggested that out of these organisations such a Federation might well be created. What is wanted is a powerful national organisation of a form that will secure the general confidence of farmers, labourers, the distributors of all classes and also the support of those who are employed in the various processes of conversion of produce into food, the workers in the mills, the bacon, jam and sugar factories, and so forth. The creation of such a representative organisation is an essential of success. It may be necessary in its earlier stages to limit its powers; indeed, it may have to start as a purely advisory body. But even then it could, in conjunction with the National Agricultural Authority, help to provide the directing force needed for effective action. This is no new idea; there is an admirable Federation of this character in Sweden, and others, I understand, though I have no personal knowledge, exist in Denmark and other countries.

Ultimately such a Federation might be put into a position to take definite responsibility for securing that land is cultivated to the best advantage, for discouraging the accumulation of surpluses in home production, and for the putting and keeping the marketing side of agriculture into order. But it should refrain from interfering, except in cases of special urgency, with the productive work of farmers; for farmers are and should be individualists and must not, under any circumstances, be harassed by undue control. The late Lord Milner in his series of Essays published under the title of *Questions of the Hour*, discussed this particular problem with his usual profound understanding, and referred to the essential advantages of this method of action. "It places," he says in an illuminating phrase, "the onus of improving the conditions of any industry upon the people actually engaged in it, instead of attempting to effect such improvement by external pressure."

The essence of administrative success is therefore the appointment of a National Agricultural Authority to represent the national interests and the creation of a National Federation of Agriculture to represent agricultural interests. It is eminently desirable to give these two authorities, either separately or conjointly, the widest possible powers of action. This is the only way of escaping from the disastrous influence of political control and getting on with the work of reconstruction.

A similar line of administrative policy is also necessary for the various branches of industry if we are to secure practical results.

PART V

THE RIDDLE OF CIVILISATION

PART V

THE RIDDLE OF CIVILISATION

1. SOME GENERAL REMARKS

It must not be thought that I am so foolish as to suppose that the proposals already set out give an answer to the riddle of civilisation. They are only a first step towards the solution of Britain's immediate problem. They constitute, however, an important step, for if we take care to secure that the development of industry and agriculture go together, and that industrialists and agriculturists expend their increased wealth for mutual benefit—do their shopping at home and not abroad—it will enable us to absorb a large part, possibly as many as two millions, of our unemployed. We shall also be creating the wealth needed by the nation for other purposes. But we cannot stop there; we have to go on with the building, with a clear plan and a conscious effort. We have to utilise some part at least of the material wealth so created to deal with the housing problem (91) and the further development of agriculture.

But what is all-important in this line of policy is not that it will reduce unemployment and create wealth. It is something quite different; it is this.

When we have made Britain as self-supporting a nation as possible, with overseas trade not our primary interest, but only the second string to our bow, we shall have brought the whole social problem within our control, and its political, economic and financial aspects can be dealt with together. We shall no longer, to use a phrase of Mr. J. A. Hobson, be "at the mercy of great economic causes over which we have no control." We shall have got control; we shall be masters of our own fate. Concurrently, such is the power of production that rationalisation gives us, we shall find ourselves producing much more wealth than we can consume. But having control of the situation we shall be in a different position from what we are in to-day. Safeguarded from the evils of outside competition it will become possible to raise rates of wages and so the standard of living; thus we shall increase consumption and bring it more into accord with production. It is true that we shall not yet have solved the problem of unemployment, for rationalisation, unless it is controlled, will tend to reduce employment once more. Nevertheless, when we are ourselves providing the greater part of the wealth needed for our civilisation, and we have control, the problem of unemployment assumes a different aspect; it becomes a problem of absorbing our surplus energy in purposes valuable to our civilisation. It becomes a problem of the utilisation of leisure.

2. THE FLOWERS OF NARCISSUS

We can then, having got so far, go to another source for the solution of the riddle of civilisation. "If," said Mahomet, "a man finds himself with loaves of bread in both hands, let him change one for flowers of Narcissus."

Having created the necessities of civilisation, we can then turn our attention to the flowers of civilisation. Our surplus energy, represented by the unemployed, can be utilised in the developing of knowledge, the securing of a higher standard of health, the creation of leisure by the shortening of the hours of labour and the general reduction of the strain of life. We can develop the arts and the crafts, and in agriculture the fascinating and health-giving occupation of gardening; we shall have more time to cultivate the drama, music and literature, and also for travel.

Thus we shall cultivate the gentle art of happiness—the flower of civilisation.

3. CONCLUSION

I do not propose to close this essay on a somewhat sentimental note, which may leave my readers dreaming, but with a more practical comment.

We must, to deal successfully with the problems of the day, clear our minds not only of Utopian dreams of the future, but of the abstract theories of the past. We must adopt a scientific method of thought; we must learn the lessons of history; we must study the facts and the solutions that these facts suggest. Even an elementary study of the facts will bring out one important point: almost all, perhaps all, the problems that we have to deal with in our national life have already arisen in this or other countries and civilisations, and have been dealt with successfully, or if not successfully, yet in such a way as to show exactly where both the difficulties and the solutions lie.

To secure the Rebuilding of Britain there are then four things to be cultivated: suspicion of abstract theories; knowledge of facts; the art of constructive thinking and the habit of getting things done.

APPENDIX

NOTES ON INTRODUCTION

1. The Dealers' Charta, 1844: 7 & 8 Victoria, Cap. XXIV, was the last of a series of statutes passed to secure "free trade" in commodities and to legalise the dealers' position. It treats of internal trade only.

2. The story of this controversy is very well explained in contemporary pamphlets by W. Illingworth, 1800 (British Museum Catalogue 518, h. 19). It is also dealt with in the author's work entitled *The Rebuilding of Rural England*, and Mr. Penty's book on *Protection and the Social Problem*, referred to above.

3. Mrs. Sturge Gretton, in her work entitled *Some English Rural Problems*, states, however, that some prosecutions for dealing in food supplies continued during the first quarter of the XIXth century.

4. The practice of giving family allowances out of parish funds appears to have been first instituted, without special legislation, at Speenhamland, in Buckinghamshire, in 1795. The custom spread throughout England in the early years of the XIXth century.

5. The Anti-Corn Law League, founded in 1838, had as one of its objects "to teach the agriculturalist that 'they' (the Corn Laws) had not even the solitary merit of securing a fixed price for food" between nations. Cobden developed the point and claimed that free trade would stabilise prices.

6. These phrases are from a speech delivered by John Bright at Birmingham in 1864. See *The Speeches of John Bright*, edited by Thorold Rogers, Vol. ii, p. 358, a book presented to me by the Cobden Club for my prize essay on free trade.

7. The Poor Law Amendment Act, 1834, 4 & 5 William IV, c. 76, put an end to the system of family allowances.

8. There can be little doubt that it was the Irish famine that created the phrase "The Hungry Forties."

9. An admirable analysis of Wheat and Bread prices, with the relative import duties from the years 1800 to 1913, appears

in *The Miller*, February 3rd, 1913; from it these figures are taken.

10. Mr. Adam Smith's exact view on trade between different countries appears in *The Wealth of Nations*, Book IV, Chapter I. In dealing there with the "Benefits a nation derives from foreign trade," he states that "Between whatever places foreign trade is carried on they (i.e. the nations) all of them derive two distinct benefits from it." "It carries out that surplus part of the produce of their land and labour for which there is no demand among them and brings back in return for it something else for which there is a demand . . .; they all receive great benefit from it." These phrases give an interesting, if highly imaginative, picture of what trade might be, not, of course, what it is. It is important to realise quite clearly that *The Wealth of Nations*, an extraordinarily interesting work, is largely imaginative, and that many of the statements and the theories are not based on facts—which at that time had not been investigated.

Sir William Beveridge, President of the London School of Economics, a defender in *The Times* of March 25th, 1931, of free trade, in the course of a long letter on the subject states the free trade case as follows: "The general free trade argument is that international trade being exchange, imports and exports of goods and services are used to pay for one another." He makes certain exceptions to this rule, but he does not seem to differ in essentials from Adam Smith. The letter may be referred to.

11. It was not only Cobden who advocated free trade as a means of developing Agriculture. Francis Quesnay (1694 to 1774), who may be considered the creator of the free trade idea in its modern form, advocated free trade in order to promote the interests of agriculture by increasing the price of grain. See *Protection and the Social Problem*, by Arthur J. Penty.

12. The opinion of leading authorities as to what proportion of our food supplies could be advantageously produced at home, with references, is given in a pamphlet entitled *Britain's Food Supplies*, issued by the Rural Reconstruction Association, Leplay House, 65, Belgrave Road, London, S.W. It appears clear that Britain could without difficulty produce at least 80 per cent. of its own food requirements.

13. Opposition to the free trade theory appeared in the

Labour movement in the early years of the present century. After the War, this school of thought became more influential; it secured support in the Independent Labour Party, and the late Mr. John Wheatley, who was the Minister of Health when the Labour Party was in power in 1924, was a keen opponent of free trade theory. More recently, notwithstanding that the official view of the party remained in favour of free trade, the opposition grew. Mr. John Beard, President of the Trade Union Council, in his presidential address, delivered in September 1930, gave a reasoned criticism of the party's attitude to the question; he declared that the antithesis to free trade was not import duties but regulated trade, and recommended the Labour Party to adopt that point of view. The Rt. Hon. Christopher Addison, Minister of Agriculture in the Labour Government of 1929-31, also criticised acutely free trade policy, and in a speech delivered to his constituents on August 23rd, 1930, declared that free trade in its old interpretation was "as dead as Queen Anne" (see *The Times*, August 25th, 1930). Nevertheless, at the general election in 1931 the Labour Party generally declared itself for free trade, though many and probably most candidates qualified their support, and put forward concurrently anti free trade proposals, especially for dealing with agriculture. The New Party, led by Sir Oswald Mosley, formed by Members of Parliament and others who broke off from the Labour Party early in 1931, are anti free traders.

14. Mr. Sidney Webb's opinion on free trade is given in the year 1923 by Lord Milner in his book *Questions of the Hour*. Mr. Webb at that time held that it was wrong to regard international trade as a struggle between nations because such trade is mutually beneficial. It was this theory that he taught to the Labour Party for a quarter of a century. Since then he may have somewhat changed his point of view.

15. Karl Marx, in the course of a speech delivered before the Democratic Club in Brussels on January 9th, 1848, said: "The free trade system works destructively. It breaks up old nationalities and carries antagonism of proletariat and bourgoisie to the uttermost limit. In a word, the free trade system hastens the social revolution." Referred to in a letter in the *Morning Post* of July 18th, 1929.

16. See an article by Mr. Snowden in the *Daily Herald*, August 8th, 1927.

17. According to a report which appeared in *The Times* of October 22nd, 1930, Mr. Snowden, speaking at Accrington on the previous day, had remarked: "It has been said that if you could induce the Chinese to wear shirts of British cotton goods only two inches longer, there would not be enough spindles and looms in Lancashire to supply the extra demand in normal times." See also correspondence in *The Times* of October 24th, 1930.

18. At the time that Mr. Snowden was delivering his address, a British Economic Mission, sent by the Labour Party, was on its way to the Far East to investigate conditions. Its report, issued by the Department of Overseas Trade, in April 1931, does not give support to Mr. Snowden's hopes. The cotton trade, it appeared, had been in the main captured by the Japanese, and the only possibility of recapturing it was to initiate a trade war, based on lowering prices.

NOTES ON PART I

Trade in Theory and Fact

19. See *The Times*, April 5th, 1931.

20. Professor Stanley Jevons, in his work on *The Coal Question*, published so long ago as 1866, emphasises and clarifies the position. Coal, he explains, is capital which is being used up, and not a crop that can be re-created annually. Whilst there is little danger of our coal being exhausted it can only be obtained at a steadily increasing cost. Moreover, it may be pointed out, coal-mining is hardly a desirable occupation for civilised man, and it should therefore be reduced to the absolute minimum.

21. The best way of dealing with the problems that will ultimately arise from the importation of timber is to proceed with afforestation in our own country. Dr. Schlich, in his work on *Forestry in the United Kingdom*, estimates that between five and six million acres can be brought under afforestation, give

temporary employment to 500,000 men and permanent employ to about 100,000. Such development would be followed by an increase of national wealth.

22. The following information throws light on the question here discussed as to what proportion of our imports could be produced in our own country.

The imports for 1930 and 1931 respectively were as follows:

	1930	1931
Food, drink and tobacco ..	£475,116,083	£416,999,105
Raw materials, etc.	250,458,815	173,366,726
Manufactured articles, etc. ..	307,417,875	261,972,398
Miscellaneous and parcel post consignments	10,982,488	9,836,480
Total ..	£1,043,975,261	£862,174,709

Sir Charles Fielding, in his booklet *Your Road to Prosperity*, estimates that we could, to replace imports, provide with advantage from the land of this country produce to the value of £382,000,000.

The Empire Industries Association has made careful investigations on the subject of the articles wholly or mainly manufactured in foreign countries that come into this country. They estimate, on the basis of the figures for 1929, that articles of a value between £205,000,000 and £250,000,000 could be advantageously produced in this country.

Taking these estimates, we arrive at a total of between £637,000,000 and £682,000,000. It is therefore safe to assume that we could advantageously produce at home £500,000,000 worth of the food, manufactured goods and other articles imported. We actually imported cotton goods in the year 1931 of the value of about £9,000,000.

The importance of home trade, even as it exists to-day, is often overlooked; it is stated by some authorities that in the year 1929, 72 per cent. of the total output of collieries, factories and agriculture of Britain was consumed at home. It is suggested that 90 per cent. of our requirements could be advantageously produced at home.

23. Under the Treaty of Shimonoseki the Japanese secured for themselves the right to set up factories in Chinese Treaty

Ports and in Central China; further, the goods produced by these factories, though apparently subjected to a tax equal to those on imported goods, have exemption or special preference in likien duties and internal taxation. See *A Son of China*, by Sheng Chang. Japanese investors have taken advantage of these rights and privileges to establish large numbers of admirably equipped cotton factories in China, directed by Japanese and other foreign specialists. The Chinese employees in these factories draw pay at rates said to be about one-tenth of English rates. See *The Times*, February 23rd, 1931.

24. Reports from Belfast received in 1931 went to show that it was then hoped by rationalisation of the linen industry to undercut in price certain sections of the cotton trade and so secure increase in the use of linen. Any success so obtained will, there can be little doubt, cause unemployment in the cotton industry.

25. The imports of wheat from Russia in 1930 amounted to 18,717,260 cwts., representing 18 per cent. of the total imports into Britain from all countries. The corresponding figure for 1931 was 28,934,088 cwts., 24 per cent. of the total.

26. It is now known that prices are not fixed in accordance with supply and demand, and, in fact, there is no such thing as a "law of supply and demand." The phrase, like "free trade," merely confuses the issue. (See pp. 70 and 119).

27. Mr. Scullin, when Labour Prime Minister of Australia in 1931, and Mr. Fenton, Minister of Trade and Commerce in his government, were constantly emphasising the importance of making Australia as self-supporting as possible, and it does not appear that his political opponents differ from him on this point. British firms were in 1931 extending their factories and building or contemplating building new ones.

28. *The Observer* of February 22nd, 1931, stated on the authority of Mr. Percy Sparkes, a representative of the Textile Manufacturers of Canada, that three British woollen factories had recently been opened in Canada, and many more were contemplated. It appears probable that by the end of 1932 all Canada's requirements of woollen cloths, with the exception of certain specialities, will be provided for by Canadian factories.

29. The facts relating to overseas investments are dealt with

from time to time in *The Economic Journal*, the *Board of Trade Journal* and *The Economist*. The nominal value of these investments appears to be about £4,000,000,000, and the net marketable value in the first half of 1931 about £3,000,000,000. The average net income for the 5 years 1925–30 may be computed at about £230,000,000. These figures are not given for quotation but to give a rough picture of the situation: there are many minor problems involved. These figures may vary with the value of the pound sterling. In the past some part of this income was reinvested abroad; but the net balance comes into this country in the form of goods.

The importations representing interest and dividends on investments do not, it may be noted, necessarily come from the countries where the funds are invested, but may come through other trade channels.

The so-called "balance of trade" is arrived at by adding to the value of exports the value of the tribute trade and certain other items, such as shipping and insurance, and comparing the total with the value of imports. In this relation it should be noted that there is due to us on what is really an account for exports never paid for, a sum of about £3,000,000,000 represented by foreign investments, and whilst that remains unpaid the theory of the balance of trade is tainted by fallacy.

30. This slight description of the Slavonic workers in Denmark was given to me by a Danish lady in the course of a description of her life as a child in a Danish provincial town.

31. Danish exportation of bacon is dealt with in the Appendix to a booklet entitled *The Organisation of Marketing in British Agriculture*, issued by the Rural Reconstruction Association.

32. What will be for Britain a new form of trade, arising from the repayment of money invested abroad, may become a very important element in trade of the future. It might cause much dislocation in British agriculture and industry, and wholesale unemployment, unless special measures are taken to limit it to such products as this country is not able advantageously to produce itself. So limited, it might be of great national value.

33. The indirect control referred to in the text is dealt with in my book entitled *A Short History of English Rural Life*, pp. 111, 2. George Allen & Unwin.

34. Further information relating to the historic incidents referred to in the text is given in my book, *The Rebuilding of Rural England*, referred to above.

35. It has been stated that 100 country mills were shut down in the decade between 1920 and 1930.

36. Wheat prices varied in the XIXth century between about 18s. and 180s. a quarter; in the present century between about 20s. and 95s. In the year 1931 the price fluctuated between about 20s. and 31s. a quarter, and it varied every hour and in every market.

37. See *Prosperity for England*, a pamphlet by Sir Charles Fielding.

38. See *The Bread of Britain*, by A. H. Hurst, pp. 36 and 37, and also 50 and 51.

39. At the Conference of the Amalgamated Union of Operative Bakers, Confectioners and Allied Workers held at Portsmouth in August 1930, Mr. H. G. Keens, the President of the Union, computed the cost to the householders of this country of a rise of a penny on the quartern loaf at £9,000,000 a year.

40. See *Principles of Political Economy*, by Charles Gide.

41. The annual imports of bacon, pork, ham and other pig products averaged for the five years 1925-30 about £60,000,000, and the corresponding figures for home production is estimated at about £24,000,000. The best available estimate of the number of additional workers who might be employed in the production of pigs, bacon factories and relative occupations in this country is 120,000, but it may be far more.

41a. This story admirably illustrates the position to-day, when a very large number of cultivators are in the hands of rural financiers.

42. The result of the investigation made by Sir Charles Fielding on the cost of distribution appears in his book entitled *Food*, published in 1923. His enquiries go to show that at that time between 50 and 60 per cent. of the retail prices of home-grown wheat, milk and meat was absorbed in the process of distribution and conversion into food, and the corresponding waste involved by our present system so far as these special products were concerned was about £135,000,000 a year. If this be correct—and there is no reason to doubt its substantial

accuracy—the total waste in distribution and conversion of food products is probably over £200,000,000. See also *A National Rural Policy*.

The results of admittedly incomplete investigations that I made some years ago were published in my book, *The Rebuilding of Rural England*, referred to above. It appeared probable that the costs of distribution and conversion into food of home-grown products averaged somewhere about 56 per cent. of the retail price; whilst rent, rates, taxes and interest on capital absorbed perhaps 10 per cent.; this would leave to the farmer and labourer for their productive services about 34 per cent. of the retail price.

It is thought that the rationalisation of the work of distribution and conversion might reduce the average percentage costs from 56 per cent. to somewhere about 36 per cent.

Such figures can only be rough estimates, but they are near enough to the truth to give a fairly accurate picture of the position.

An exhaustive investigation of this aspect of our social problem is needed.

NOTES ON PART II

Some Effects of the Trade War

43. The problems of rationalisation arose in the first instance from the introduction of machinery which replaced workers by hand. These problems are dealt with in the first half of the last century in the writings of Robert Owen and his contemporaries, who foresaw the difficulties that are facing us to-day.

44. The phrase "trade cycles" explains nothing; such cycles are associated with variations in price levels and increase and decrease of trade, but which is the cause and which the effect is not made clear by the authorities, so far as I have examined them. It should be realised that such cycles are social evils that can be dealt with by a change of national policy.

45. The Prince of Wales, on his return from South America in the spring of 1931, whilst urging British manufacturers to

do their utmost to extend their trade in South America, pointed out that efficient local industries were growing up in nearly all the South American Republics protected by tariff walls which blocked the importation of competing goods from Britain. See *The Times*, May 22nd, 1931. If tariff walls are found to be unsuccessful there can be little doubt that further steps will be taken to protect home industries. Prohibition of imports is already taking the place of tariffs; indeed, the embargo on the sale of British cotton goods in India may well become a model for action by other countries.

46. There will be also a tendency for manufacturers from other countries where wage rates are high, such as the United States of America, to set up factories in Britain, but this can hardly go on on a very large scale.

47. Mr. Cyril B. Dallow, a Liberal candidate at the General Election of 1931, stated in his address, "the superiority of our Workmanship and the enterprise and astuteness of our Traders, which together built up our great commerce of the past, if now given the same Freedom of Trade conditions, will soon restore the impaired fortunes of our Country, provided peace is secured at home."

48. The fact that overseas trade provides occupation for people employed in shipbuilding and the actual transporting of goods across the seas is not overlooked, and although it seems a strange defence of trade that moving things about the world gives occupation, the facts must be faced, and it must be admitted that if we develop home production, employment in the shipping industry will be reduced. Fortunately, this reduction of employment will not be large. So far as I can ascertain, the number of persons of British nationality employed at the end of 1931 in the shipbuilding industry, the overseas trade and the loading and unloading of ships can hardly be over 300,000. A policy of national development might put half of these out of employ. On the other hand, the increase of purchasing power that would follow the development of our own country might increase the importation on a considerable scale of such raw materials as cotton, and possibly of articles of luxury and of goods that we cannot produce at home: Moreover, our fishing trade might be developed. All this would increase employment in the shipping industry.

Measured in tonnage, it appears that about 40 per cent. of our overseas trade comes and goes in foreign ships and this proportion might be reduced to our national benefit. Information can be obtained from the *Statistical Abstract*, the *Balfour Reports*, particularly the last report Cmd 3282, the *Survey of Overseas Markets*, the *Board of Trade Journal* and the *Ministry of Labour Gazette*.

49. See *Parliamentary Reports*, February 16th, 1931, Vol. 248, col. 909.

50. See the *Daily Telegraph*, January 21st, 1931.

51. This information was given by a speaker at a Conference organised by the National Industrial Alliance of Employers and Employed at Balliol College, Oxford, on July 19th and 20th, 1930.

52. In Brazil, surplus coffee was in 1931 actually burned. See *The Times*, April 5th, 1931.

53. Recent variations in price levels, combined with want of exact official estimates, make it difficult to give correct up-to-date estimates of the productivity of Britain. The Census of Production published in 1925 gives the value of the agricultural output of Great Britain at about £285,000,000 a year, and even when adjustments are made for reduced production and lowered prices the value of the output in the year 1931 could hardly have been below £200,000,000. The number of persons of all classes regularly occupied in agriculture in the year 1931 was about 1,200,000. It is admitted by even the least sanguine authorities that we could double production; this should absorb about a million workers. The figure of £250,000,000 which appears in the text is based on these estimates, rearranged as follows:

Under-production of National Wealth £200,000,000
Cost of Unemployment Insurance at the rate
  of £50 per worker 50,000,000

                                    Total .. £250,000,000

NOTES ON PART III

VARIOUS MATTERS FOR CONSIDERATION

54. The point referred to in the text of a political party running two opposing theories was a conspicuous feature of the general election of 1931, when some and perhaps many Labour candidates declared themselves in favour of free trade, i.e. free imports, and the antithesis to free trade, i.e. control of imports by the special methods of import boards.

55. This computation of the world surplus of wheat was given by Sir Daniel Hall, the well-known agricultural authority, in an address over the wireless on May 18th, 1931.

56. The information relating to the variation of prices of grapes in Covent Garden is given on the authority of Messrs. J. Lyons & Co., Ltd., the well-known catering firm.

57. In the *Financial and Commercial Review*, issued by *The Times*, on February 10th, 1931, appears a chart of the variation of the level of commodity prices from 1783 up to 1930, which may be referred to. These variations are undoubtedly largely due to the changes of money values.

58. The "world's price" at one time, by a strange inversion of ideas, secured for itself the title of the "economic price"; thus arose endless confusion of thought. The phrase "economic price" is in modern parlance properly used to describe the price that gives the best economic results.

59. The price system, so far as it affects agriculture, is dealt with shortly but clearly in *The Report and Declaration of Agricultural Policy adopted by the Policy Committee of the Central Chamber of Agriculture in January and February*, 1931; also in the publications of the Rural Reconstruction Association, and in my book, *The Rebuilding of Rural England*, referred to above.

60. Recent reports of the Ministry of Health go to show that the average number of men, women and children in receipt of Poor Law Relief in 1931 was about 1,000,000. What proportion of these recipients would under normal conditions be employed in industry or agriculture is not clear. See a speech by Major Elliot in the House of Commons of April 14th, 1931, in *Parliamentary Reports* of that date.

61. See Report of Ministry of Labour, entitled *Report on an Investigation into the Employment and Insurance History of a sample of persons insured against Unemployment in Great Britain*, pp. 45–48.

62. These figures are given on the authority of the late Mr. Wm. Graham, President of the Board of Trade in the last Labour Government. See *The Times*, April 10th, 1931.

63. This information is given on the authority of a statement made in a book entitled *A Corner in Gold*, and is discussed in Mr. Arthur Kitson's work entitled *A Fraudulent Standard*, pp. 153–155.

64. See Professor Lorias's *Economic Foundations of Society*.

65. One indirect result of variations of prices and of value of money, the importance of which is often overlooked, is to undermine the reality of arguments based on prices and also to render statistical calculations based on money values unreliable. Such arguments and calculations constantly lead to entirely misleading conclusions.

66. The National Confederation of Employers' Organisations, in their evidence before the Royal Commission on Unemployment Insurance, submitted that the figures showed that one-third of the whole population derived benefits from the payments for unemployment and social services. See Report of the Commission when issued, and *The Times*, May 6th and 7th, 1931. The proportion may be higher.

67. The National Government formed in August 1931, from the members of all parties did, in fact, under the special conditions created by a national emergency, raise taxation and reduce the salaries of teachers in schools receiving government grants, civil servants and the Army, Navy and Air Forces, also rates of payments made to the unemployed. But the strong feeling so caused may deter future governments from any similar step. It remains to be seen whether their action will secure the results hoped for. It is possible that the taxation will not, in fact, balance the budget, whilst the reductions of incomes of all classes will result in increase of unemployment.

68. There is another somewhat similar problem involved from the creation of the various war loans and other national and local non-wealth-producing investments. Here also we collect funds in order to provide interest on the various loans

from one group of individuals and transfer it to another. Here again we are involved in a situation from which we cannot escape; and here again the increase of national wealth will help us to bear the burden of our obligations.

69. See *The Times*, February 24th, 1931.

70. See *The Times*, March 3rd, 1931.

71. The embargo by the Canadian Government on the importation of Danish bacon in 1931 was made on the ground of danger of importation of disease; but was that the real reason? A Canadian speaker at a meeting held in London by the Young Farmers' Clubs in 1930 explained that Canadian policy was directed to developing the pig trade and capturing the British markets.

72. The phrase "Public Utility Society" is used in England to describe a special form of organisation in which profits are limited. In America the same phrase has a different meaning: in the text the words are used in their English sense.

73. The Co-operative movement in all its various forms is not dealt with here, as it would require very lengthy description; it can be studied from the numerous works on the subject. Notwithstanding the fact that it is imbued by free trade ideas, it is essentially an anti free trade movement.

74. "Socialism" and "Safety First" were the slogans adopted by the Labour and Conservative parties respectively in the General Election of 1929.

75. It should be noted that there is a school of thinkers who might deem themselves internationalists who are definitely advocating the regulation of international trade, and in that sense are anti free traders.

76. In the course of the debate that took place on the occasion of the National Conference on Agriculture held in February 1929, representatives of the National Union of Agricultural Workers argued that in order to improve the condition of agricultural workers in this country we must look to the League of Nations or some other international organisation to raise labour standards and conditions in other countries. The discussion that took place on this point at the Conference is of considerable interest; it illustrated the various points of view. See *Report on the National Conference on Agriculture, February 25th and 26th, 1929*, issued by the Central Chamber of Agriculture.

77. The fact that economics has to some extent become a dogma has had a serious effect on the position of some of the younger men who are taking up the teaching of economics as a career. "I could not say what you say," a brilliant young economist remarked to me a few years ago. "I must be a free trader, otherwise I should never receive promotion." He was a young man with a wife and two charming children to support.

78. "Empire Free Trade," if it has any exact meaning, suggests, if we are to accept Lord Beaverbrook's explanation, free trade within a tariff boundary. Free trade within a boundary was adopted in Britain in the early years of the last century, and its exact effect in recent years may be studied in the United States of America. In both cases the policy was accompanied by disaster. On the other hand, there is much to be said for making the Empire an economic unit with a uniform currency and a controlled financial system. We could thus deal, so far as our colonies are concerned, with people anxious to secure a rising standard of life and more likely to pay a fair price for what we have to sell. If this line of policy were adopted, free trade could be definitely abolished within the Empire and complementary trade organised. This would accord with the views of the colonies, so far as can be ascertained. Probably that is what the more intelligent Empire Free Traders have in mind—the antithesis to free trade. Such confusion of thought is characteristic of modern politics.

NOTES ON PART IV

Rebuilding Britain

79. The National Government instituted, in the autumn of 1931, a national campaign urging everyone to "Buy British," i.e. to buy home or Empire produce in preference to foreign.

80. See Report of the Royal Commission on Unemployment Insurance when issued and *The Times*, May 5th, 1931.

81. See *The Financial Policy of Czecho-Slovakia*, by Alois Rašín.

82. See *The Mind and Face of Bolshevism*, by René Fülöp-Miller.

83. Constructive thinkers of a distinctive school of thought have been at work on the problems of reconstruction since the early years of the present century. Two modern movements were based on the theories of this school, the first, the National Guilds League, which advocated during its short career important proposals for reconstructing industry; the second, the Rural Reconstruction Association, that deals with the reconstruction of Agriculture. Adoption in December 1931 by the National Government of the so-called Quota System for wheat, is perhaps the most striking example of the influence of the new point of view. The Quota policy so laid down for wheat, really the antithesis to free trade, has general application to agriculture and industry. It has long been advocated by the modern school of thinkers referred to above. This line of policy was constantly discussed in the year 1931 in the press, and, so far as it refers to agriculture, has been endorsed by the Central Chamber of Agriculture and the Scottish Chamber of Agriculture in recent reports. One hears vague talk of a new political movement with the title of "Tory Socialism" formed to support this policy.

84. This analysis of the relation of agriculture to unemployment, though sufficiently accurate to give a picture of the position, requires qualification and explanation. It is quite accurately pointed out in the text that if agricultural produce comes into this country to pay the interest and dividends on foreign investments, it is liable to put two groups of workers, agriculturists and industrialists, out of employ. On the other hand, the individuals who actually receive the interest and dividends may spend their receipts at home in such a way as to increase employment in this country, or, on the other hand, they may spend or reinvest it abroad, as no doubt often occurs, and not increase employment at home. It all depends on how the money is spent. Home production will also be accompanied by a loss of shipping trade referred to above (see page 209).

85. It is constantly assumed that the climate of Britain is unsuited for agricultural development. Uncertainty of climate is always the enemy of agriculture, but Britain has the great advantage of a moist, temperate climate without, save on rare

occasions, great heat or cold. The destruction of a corn crop by spring frosts, droughts, excessive heat or other similar disasters, such as may occur in America, India or Russia, are substantially unknown in Britain. What the British farmer needs is better drainage, a complete system of irrigation and the introduction, through electricity or otherwise, of arrangements for drying his hay and wheat crops in the comparatively rare cases of wet harvests. Moreover, it should be noted that stock in all its forms, which may well become the basis of British agriculture, is not affected by weather in the same way as corn crops.

86. The law of increasing returns is constantly confused with the law of diminishing returns, which mid-Victorian economists believed had at that time special application to agriculture. It is clear that under present conditions we could very largely increase production with reduced relative costs. It is no doubt true that at last we reach a point, in all branches of agriculture, when increased production is only possible with increased relative costs. See *A National Rural Policy*, pp. 17 & 18.

87. The opinion referred to in the text was given in a speech delivered at King's Lynn on March 14th, 1931, and reported in *The Times* on the following day.

88. The support of the Prime Minister is recorded in his exact words in a leaflet issued by the Rural Reconstruction Association.

89. The Agricultural Marketing Act, 1931, Clause 15, authorises the appointment of Commissioners with power to investigate the problem of distribution of agricultural produce and to prepare schemes in conjunction with the present distributors for its organisation.

The policy recommended in the text can therefore be started on at once by any Minister of Agriculture who wishes to secure a home market for the British farmers. He would have, of course, to secure financial support for his schemes.

90. See the Report of the Central Chamber of Agriculture for the year 1930, and subsequent discussions on the problem reported in the *Journal* of the Chamber for 1931.

NOTE ON PART V

THE RIDDLE OF CIVILISATION

91. The housing problem involved the provision of many hundreds of thousands of homes for our working people. There are two obstacles to be faced as soon as any attempt is made to approach this problem. First, we have yet to decide where it is desirable to provide homes for the workers, and this question in its turn depends on our national policy. If we are expecting to develop industry on traditional lines, no doubt our workers must live in or near the present industrial towns. If we are going to base a new national life on agriculture and decentralised industry, we have to develop our country towns and villages or build new towns in the centre of rural areas. Secondly comes the financial problem: How are we to build houses for working people to be let at rents that workers of the present generation can afford to pay? This problem remains unsolved, but specific proposals have been put forward. It is suggested that all rates and taxes, direct and indirect, should be removed from new houses and that the cost of building should be financed by the use of what is now called "fiduciary" credit or currency. Such proposals, which are of the nature of a reversion to early methods, however good, cannot be expected to be accepted in the present state of public opinion and may therefore be ruled out for consideration in another decade. See *A National Rural Policy*, where some interesting information on this subject appears.

INDEX

This index is intended to supplement the Table of Contents, which should be referred to.

Books and documents referred to are indexed under the authors' or issuers' names, if known, and the titles are printed in italics.

Trades and Industries, specific and national, are indexed under the heading of the specific article or the nation concerned.

INDEX

This index is intended to supplement the Table of Contents, which should be referred to.

Books and documents referred to are indexed under the authors' or issuers' names, if known, and the titles are printed in italics.

Trades and Industries, specific and national, are indexed under the heading of the specific article or the nation concerned.

Abadan, 99

"Act of faith," 25, 32, 73, 132, 155.

Addison, Dr. C., 179, 201.

Advertisements, 64, 94, 105 f., 113; and the Press, 105 f.

Afforestation, 175, 180, 202 f.

Africa and trade of, 32.

Agricultural Authority, 188; a National, 189, 191.

Agricultural, Federation, 189 f.; labourers, see workers; Marketing Act, 1931, 144, 189, 215; National Conference, 186 f., 213; production, see Home-production; workers, 28, 42, 46, 108 ff., 131, 172, 175, 179, 212.

Agriculture—business side analysed, 68 ff.; cost of under-development of, 112, 173 f., 210; Minister and Ministry of, see Minister, etc.; productivity of, 111; and Unemployment, 42, 46 f., 53, 57 f., 74, 93, 110, 112, 129 f., 171 ff., 183, 214 f.

Allotments, 177.

Allowances, family, 24, 29, 199.

Almonds, 41.

Amalgamated Union of Operative Bakers, etc., 206

America (United States of) and trade and finance of, 21, 38 f., 50, 99, 103, 135, 137, 143, 208, 212 f., 215.

Anglo-Persian Oil Company, 99.

Anglo-Saxon workers, 99.

Anti Corn Law League, 199.

Apples, 111.

Argentine, The, and trade of, 50, 103, 143.

Armaments, trade in, 37, 44.

Arts and crafts, 64 f., 197.

Australia—and trade of, 50 f., 103, 143; trade policy of, 51, 204.

Austria, finance of, 138.

Bacon—and trade in, 49, 133, 145, 207; Danish, 49, 55 ff., 81, 110, 144 f., 206, 212; factories, 82, 85, 175.

Bakers and baking, 69 ff., 177.

"Balance of Trade," 205.

Balfour Reports (Shipping), 209.

Balkans and trade of, 143.

Ballyconneely peninsula, 84.

Baltic States and trade of, 103, 143.

Bank of England, 135.

"Bargaining power," 76, 78 f., 81, 88, 120 f., 132.

"Barter Trade," 38 ff., 60, 173.

Battam's case, 61.

Beard, Mr. John, 201.

Beaverbrook, Lord, 213.

Beef, 41.

Belfast, 45.

Bergson, Prof., 14 f.

Beveridge, Sir Wm., 200.

Board of Trade—211; *Journal*, 205, 209.

Bombay, 45.

Bondfield, Miss M., 100.

Boots and trade in, 45, 99, 107.

Brazil and trade of, 38 f., 209.

Bread—also trade in, 61, 69 ff., 206; specific prices of, 24, 70 ff., 199 f.

Nationalism, 165.
Nature—Law of, *see* Law; productivity of, 111.
"New Party," 201.
New York, 135.
New Zealand and trade of, 50.
Northampton, 45.
Norway and trade of, 37, 46, 181.

Offals, 69.
Oil and trade in, 41, 102.
Optimism and optimists, 149 f.
Oranges and trade in, 41.
Owen, Robert, 207.

Paper and trade in, 101.
Parliamentary Reports, 209 f.
Passfield, Lord, 32.
Penty, Arthur J., *Protection and the Social Problem*, 20, 199 f.
Persia and Persians, 99.
Pessimism and pessimists, 149 f.
Pig products, extent and cost of importation of, 82, 206.
Pigs and trade in—69, 78 ff., 111, 212; Danish, 81.
Pitt, Wm., 22, 24.
Plato, 14 f.
Poland and trade and finance of, 49, 56, 97, 133 f., 143, 167, 184 f.
Polish labour, 56.
Political Control and Parliamentarians, 186 f., 191.
Poor Law, 125, 163, 210.
Poor Law Amendment Act, 1834, 199.
Popina, 49, 82.
Postal systems, 147, 182.
Potash and trade in, 41.
Potatoes—and trade in, 75 f., 119; specific prices of, 75 f.
Poultry and trade in, 48, 111.
Press, the, *see* Journalism; advertisements, *see* Advertisements.
Price and prices. (*The prices quoted of specific articles are indexed under the articles concerned.*)

Price, "Common," 23; controlling and fixing, 60 f., 135, 146, 180 f.; "Economic," 24, 180 ff., 210; "Guaranteed," 23; "Just," 23, 124; security of, 142, 180; "Standard," 23 f., 109, 123, 181 ff.; stabilised and stabilisation of, 24 ff., 118, 144, 146, 180, 188, 199; vacillating and variation of, 62, 65 ff., 69 ff., 81, 83, 87 f., 109, 117 ff., 147, 180, 211; scales of, 181; "Worlds," 124, 181, 210.
Price controlling committees, *see* Committees.
Price system, 66, 118, 210.
Prices—low, *see* Cheapness; competition and, *see* Competition; regulated, 27, 62.
Prince of Wales, 207.
Producers' share of retail prices, *see* Distribution, costs of.
Production—over- and under-, 100 ff., 110 ff., 143, 196, 209; and productivity (of Britain), *see* Home production, etc.
Profiteering Acts, 146.
Protectionists, 24, 27 f.
Psycho-analysts, 15.
Public Utility Societies, 147, 212.

Quesnay, Francis, 200.
Quota—and Quota Systems, 144, 214; for wheat, 214.

Raisins, 41.
Rasin, Alois—166; *The Financial Policy of Czecho-Slovakia*, 213.
Rates, 84, 207.
Rationalisation, 28, 94 ff., 105 f., 129 ff., 196, 207.
Rationing of work, 131.
Relativity, 15, 20 n.
Relief works, 126.
Rent, 84, 207.
Reparations, 57, 68, 172.
Retail prices and their distribution, *see* Distribution, costs.
Rome, 52, 103.